D0571770

Popular Mechanics

BIG

IDEAS

100 MODERN INVENTIONS
[THAT HAVE TRANSFORMED OUR WORLD]

BIG

IDEAS

100 MODERN INVENTIONS
[THAT HAVE TRANSFORMED OUR WORLD]

ALEX HUTCHINSON

HEARST BOOKS
A division of Sterling Publishing Co., Inc.

New York / London
www.sterlingpublishing.com

Copyright © 2009 by Hearst Communications, Inc.

Library of Congress Cataloging-in-Publication Data

Hutchinson, Alex.

  Popular mechanics : big ideas : 100 modern inventions that have changed our lives / Alex Hutchinson.

    p. cm.

  Includes index.

  ISBN 978-1-58816-722-4

  1. Inventions--History--20th century. I. Popular mechanics. II. Title.

  T20.H88 2009

  609--dc22

2008021267

10 9 8 7 6 5 4 3 2 1

Published by Hearst Books

A Division of Sterling Publishing Co., Inc.

387 Park Avenue South, New York, NY 10016

Popular Mechanics and Hearst Books are trademarks of Hearst Communications, Inc.

www.popularmechanics.com

For information about custom editions, special sales, premium and corporate purchases, please contact Sterling Special Sales Department at 800-805-5489 or specialsales@sterlingpublishing.com.

Distributed in Canada by Sterling Publishing
C/o Canadian Manda Group, 165 Dufferin Street
Toronto, Ontario, Canada M6K 3H6

Distributed in Australia by Capricorn Link (Australia) Pty. Ltd.
P.O. Box 704, Windsor, NSW 2756 Australia

Manufactured in China

Sterling ISBN: 978-1-58816-722-4

ACKNOWLEDGMENTS

The research for this project began with an article that appeared in the December 2005 issue of *Popular Mechanics*. Numerous staff members at *PM* contributed to the success of that project; I owe particular thanks to editor-in-chief James Meigs and deputy editor Jerry Beilinson for their ongoing support and confidence, as well as to science editor Jennifer Bogo. Jacqueline Deval at Hearst Books was the driving—but, thankfully, patient—force behind this book. In addition to the panel of experts (page 11), Doug Beattie, John Beattie, and Wolf Rasmussen made significant contributions at the research stage. Moira Hutchinson performed essential work in tracking down archival documents and contemporary perspectives. Finally, I'd like to thank my parents, and Lauren King, for their patience and support throughout the project.

CONTENTS

INTRODUCTION

Tell someone that you're working on a book about the top inventions of the modern era, and you're guaranteed to start a conversation. "Really? What are the top five?" they'll ask. "Is the cordless drill on the list?" People have strong opinions, and their own experiences dictate whether they view the absorbent disposable diaper as a more significant advance than the first video game. For that matter, the same person might cast a vote for the personal computer and the Post-it note on Wednesday morning, but switch to the remote control and the pull-top beer can on Sunday afternoon.

One thing everyone can agree on is that life has changed dramatically since the invention of the transistor in 1947, which marks the beginning of the modern era for the purposes of this collection. The computers and electronics that we now take for granted can be traced back to the transistor—but the changes in the last six decades have been much broader than that. From small conveniences like Velcro and latex paint to major innovations like nuclear power and container shipping, the innovations of modern life are about more than microchips.

To assemble the final list of 100 inventions, we consulted a panel of 25 experts at 17 museums and universities across the country (see pages 11–12). Their collective expertise spanned aeronautics, biology, physics, medicine, automobiles, computers, and a host of other fields. We also tapped into the wisdom of the resident experts at *Popular Mechanics*. The result was a comprehensive list that included some obvious breakthroughs (the laser, the microwave oven), and a number of surprises. How many people realize that the inexpensive, distortion-free windows we have today are the result of a radically new method of making glass devised by Alastair Pilkington in 1952? Or that the MRI scanners found in hospitals around the world were made possible by superconducting wire first sold by Westinghouse in 1964?

Once the list was assembled, the challenge was to discover the stories behind each invention. The road from brainwave to assembly line is seldom straight, as shown by Chester Carlson's two-decade struggle to get a commercial version of his photocopier built. And the patent wars that erupted over advances like the laser, the compact disk, and the Sony Walkman give a taste of the challenge of assigning credit when fame and fortune are at stake. In the end, very few inventions spring fully formed from one person's mind—instead, discovery would seem to be an incremental process that draws inspiration from past research and current collaborators, and the same idea often strikes more than one person independently. The stories in these pages should be considered snapshots of crucial moments along that road—moments when something that once seemed impossible became possible, or something that once seemed like fantasy became real.

Alex Hutchinson

THE EXPERT PANEL

JOHN ANDERSON
Smithsonian National Air and Space Museum, Washington DC

BLAKE ANDRES
Great Lakes Science Center, Cleveland, Ohio

DENNIS BATEMAN
Carnegie Science Center, Pittsburgh, Pennsylvania

GREG BROWN
Tech Museum of Innovation, San Jose, California

STEPHEN CUTCLIFFE
Lehigh University, Bethlehem, Pennsylvania
Science, Technology and Society program

PAUL DOHERTY
Exploratorium, San Francisco, California

ROBIN DOTY
Museum of Science, Boston, Massachusetts

VICTORIA HARDEN
Office of National Institutes of Health History, Bethesda, Maryland

JENNIE HOLLIDAY
The Henry Ford Museum, Dearborn, Michigan

PETER JAKAB
Smithsonian National Air and Space Museum, Washington DC

MARILYN JOHNSON
Oregon Museum of Science and Industry, Portland, Oregon

EMLYN KOSTER
Liberty Science Center, Jersey City, New Jersey

ROGER LAUNIAS
Smithsonian National Air and Space Museum, Washington DC

SARAH LEAVITT
Office of National Institutes of Health History, Bethesda, Maryland

ANDERS LILJEHOLM
Oregon Museum of Science and Industry, Portland, Oregon

AMY LOWEN
Louisville Science Center, Louisville, Kentucky

MATILDA MCQUAID
Cooper-Hewitt, National Design Museum, New York, New York

BRAD OSGOOD
Stanford University, Palo Alto, California
Science, Technology and Society program

TREVOR PINCH
Cornell University, Ithaca, New York
Department of Science and Technology Studies

CRAIG REED
Oregon Museum of Science and Industry, Portland, Oregon

JOHN KENLY SMITH
Lehigh University, Bethlehem, Pennsylvania
Science, Technology and Society program

DAG SPICER
Computer History Museum, Mountain View, California

KATHLEEN VOGEL
Cornell University, Ithaca, New York
Department of Science and Technology Studies

DAVID WEIL
San Diego Computer Museum, California

CHERYL WOJCIECHOWSKI
Museum of Science, Boston, Massachusetts

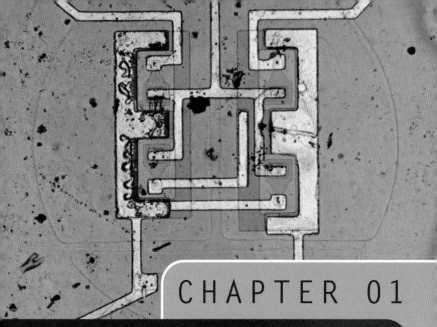

CHAPTER 01

>>> COMPUTERS

If all the inventions in this book were arranged into a family tree, the transistor would be the founding father at the top of the page. No single device has had as profound an influence on modern life, from the smallest scale—the tiny chip in a musical greeting card, now more powerful than the first room-size computers at the time of the transistor's invention—to the massive supercomputers that enable scientists to unravel the billions of sequences in the human genetic code.

The earliest computers used vacuum tubes to amplify electric signals. They were large, glass-encased, delicate, power-hungry and slow. The ENIAC (short for Electronic Numerical Integrator and Computer), a ground-breaking computer unveiled in 1946, had nearly 18,000 vacuum tubes and filled a large room. In 1939, William Shockley had first proposed the idea of replacing vacuum tubes with smaller amplifiers built from semiconductors, a class of material with the puzzling property of conducting electricity only under certain special conditions. Less than a decade later, two scientists in Shockley's team at AT&T's Bell Telephone Laboratories in New Jersey, Walter Brattain and John Bardeen, built the first "transistor"—a semiconductor amplifier that would gradually replace vacuum tubes over the following two decades.

"It is frequently said that having a more-or-less specific, practical goal in mind will degrade the quality of research. I do not believe that this is necessarily the case."

WILLIAM SHOCKLEY, Nobel Prize acceptance speech, 1956

The first transistor, built at Bell Labs in 1947, was about half an inch high and relied on two tiny gold contacts placed a few thousandths of an inch apart on a germanium crystal.

Speed and size were the transistor's two greatest strengths— and because transistors are based on the flow of infinitesimally small electrons, better technology has allowed transistors to continue getting faster and smaller. Vacuum tubes could switch on and off roughly 10,000 times a second, which put an upper limit on the speed of computer operations; the latest transistors can switch on and off 300 billion times a second, and 30 million of them could fit on the head of a pin.

The concept of "random access memory" (better known as RAM) emerged in 1956 when IBM unveiled its Model 350 Disk Storage Unit—the first disk drive. At that point, most computer data were still being stored in bins full of punch-cards, though in 1952, IBM

The first successful disk drive was IBM's Model 350 Disk Storage Unit in 1956, which introduced the concept of "random access memory," or RAM.

itself had introduced a storage system that used magnetic tape. Both systems had a basic flaw: To retrieve a particular piece of data, you had to scroll all the way through the tape (or the deck of punch cards) until you reached the right spot—a time-consuming process when dealing with large amounts of data.

The Model 350, later dubbed RAMAC, for Random Access Memory Accounting, stored its data on fifty 24-inch magnetic disks made of aluminum, each of which could rotate at 20 revolutions per second and store 100,000 characters. To retrieve a piece of data, the recording head went directly to the location where the data were stored, meaning that it took roughly the same amount of time for every request—the access was "random."

The first customer was United Airlines, which used RAMAC to set up a computerized reservations system in Denver. The RAMAC system was built with vacuum tubes, rather than the transistors that were taking over the computer industry, meaning that it quickly became obsolete: About 1,000 were built before it was discontinued in 1961. But the technology that it ushered in transformed the computer industry, allowing direct access to data and making possible the interactive computers that followed a decade later.

1959 INTEGRATED CIRCUIT

The most famous rule of the computer industry is Moore's Law, postulated by Intel co-founder Gordon Moore in 1965. He noted that the number of transistors (see page 14) that could be crammed onto a single silicon chip doubled roughly every year. In 1965, about 50 transistors could fit on a typical one-inch-diameter piece of silicon; by 1975, Moore predicted, 65,000 would squeeze into the same space. It was a bold prediction, considering that these "integrated" circuits had been invented only six years earlier.

In early digital computers, all the transistors, diodes, capacitors, and other components had to be painstakingly wired together—a time-consuming and expensive task prone to mistakes and bad connections. In 1959, two men—Jack Kilby, of Texas Instruments, and Robert Noyce, of Fairchild Semiconductor—independently arrived at a solution to this problem: All the components could be fabricated on a single piece of silicon. This integrated circuit would have many transistors etched into its surface, with lines of metal printed on top instead of cumbersome wires. Noyce was awarded the patent for the breakthrough, but both men acknowledged each other as coinventors.

"The future of integrated electronics is the future of electronics itself."

GORDON MOORE
in his 1965
paper predicting
"Moore's Law"

Moore's Law, meanwhile, has proved remarkably prescient, although the time it takes to double chip density has turned out to be 18 to 24 months rather than 12 months. Over the last two decades, observers have repeatedly predicted that the nonstop

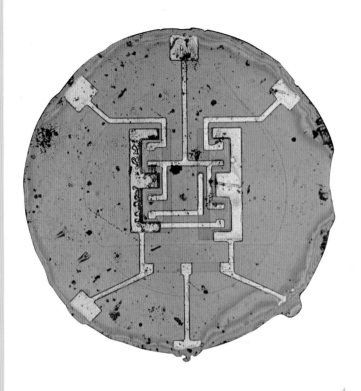

This 1961 chip from Fairchild Semiconductor—a "set/reset flip-flop" —was one of the first commercially available integrated circuits.

miniaturization will grind to a halt owing to problems such as overheating and current leakage. But so far, the integrated circuit has continued shrinking, and the latest generation contains up to 800 million transistors per chip.

In 1960, an IBM mainframe cost between $1 million and $3 million dollars; when Digital Equipment Corporation (DEC) brought out the first "minicomputer," the PDP-8, in 1965, it cost just $18,000. But the transformation sparked by DEC's PDP series was much more fundamental than just its slimmed-down size and cost. "In those days, computers were giant machines run by teams

The PDP-8 "minicomputer" included a paper-tape reader and punch for the bargain retail of $18,000.

of men in white coats," says computer historian Dag Spicer. "If you wanted to run a program, you put it on a stack of punch cards and gave it to these men. They went away, and eventually they'd come back and give you the result—assuming you didn't make any mistakes."

The precursors of the minicomputer were a series of earlier machines from DEC, which had been founded in 1957 by two MIT researchers, Ken Olsen and Harlan Anderson. In 1959, their Programmed Data Processor, or PDP-1, became the first commercial computer to feature a keyboard, monitor, and light pen (an early light-sensitive pointer that worked with cathode-ray screens) so that a single person could sit in front of the computer and program it in an interactive way. The PDP series helped to demystify computers for a generation of students who were given access to the relatively inexpensive new machines at university campuses across the country. At MIT, programmers used a PDP-1 to write the first video game, Space Wars, in 1962 (see video games, page 41). The PDP-8 model was small enough to be deployed everywhere, from submarines to the neon news display in Times Square.

Riding the wave of its minicomputer success, DEC grew an average of 30 percent per year until 1978. Its fortunes faltered when the personal computer, or PC, took over, but it was DEC's PDP computers that had introduced the interactivity that made the PC possible.

> "We thought the world would be waiting with open arms for high-speed transistor computers. But nobody cared."
>
> KEN OLSEN, on the early struggles to sell his PDP computers

At a computer conference in San Francisco in December 1968, Doug Engelbart gave a 90-minute public demonstration of the new system he and his colleagues at Stanford Research Institute's Augmentation Research Center had been developing for the previous five years. It was, in hindsight, an unprecedented peek

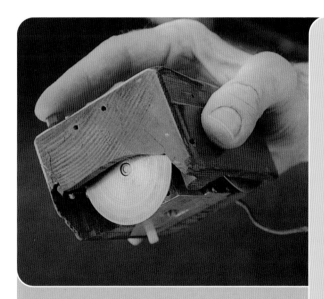

The first computer mouse, built by Doug Engelbart, was a wooden block with a single button on top. The two perpendicular wheels on the bottom were replaced in later models by a tracking ball, and more recently by laser tracking.

20 years into the future. On a 20-foot screen in front of an audience of 1,000 computer specialists, Engelbart demonstrated live video-conferencing for the first time, linking through a homemade modem to researchers back at his lab in nearby Mountain View. He also unveiled a "desktop"–style graphical user interface (GUI) based on now-familiar features like the hypertext links that allow browsers to click from page to page on the Web.

Despite these fundamental breakthroughs, Engelbart's 1968 demo is best remembered for the unassuming block of wood that he used to guide his pointer around the screen. The team had experimented with several pointing devices, but users gravitated to the "mouse," as the researchers dubbed it. The initial prototype had one button; subsequent models had three buttons that allowed users to select menus and click on objects. Engelbart wanted even more buttons, but there was room for only three of the clunky microswitches that relayed signals from the mouse buttons in those early models. At the close of the public demonstration, the audience gave Engelbart a rapturous standing ovation. But the ideas were so far ahead of their time that computermakers didn't see any potential market for them—and it wasn't until the release of the Apple Macintosh in 1984, nearly two decades later, that the mouse, along with the graphical user interface we're familiar with today, really caught on.

> "I don't know why we call it a mouse. Sometimes I apologize. It started that way, and we never did change it."
>
> DOUG ENGELBART, 1968

On October 29, 1969, computer scientists at UCLA made contact with counterparts at the Stanford Research Institute, 350 miles up the coast in Mountain View, via a dedicated phone line leased from AT&T. It was the first "host-to-host" connection on what eventually became the Internet, which at the time consisted of only those two hosts. By the end of the year, the University of California at Santa Barbara and the University of Utah had been added to the network, which was called ARPANET after its funder, the Department of Defense's Advanced Research Projects Agency (now known as DARPA). The following year, new nodes continued to be added at a rate of about one per month.

Computer scientist J. C. R. Licklider of the high-tech company Bolt Beranek and Newman sketched out the original concept of ARPANET—or, as he termed it, an "Intergalactic Computer Network"—in 1962, just before taking a senior post at ARPA that allowed him to push for its implementation. The technical breakthrough that made the network possible was "packet switching," which had been developed in the early 1960s by several groups in the United States and Great Britain working independently. In a network with many computers, building high-speed lines to connect all of them would have been impractical, and

> "The full impact of the technical changes set in motion by this project may not be understood for many years."
>
> ARPANET Completion Report, 1978

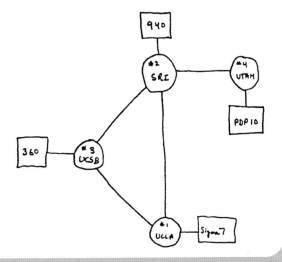

A sketch from 1969 shows the extent of ARPANET at the end of its first year of operation, with just four computers connected to what later became the Internet.

allowing a single computer to tie up a transmission line while it sent a long message would have created bottlenecks. Packet-switching broke each message into a series of small packets that were sent independently, allowing many messages to travel along the same lines simultaneously, then reassembling the packets at their destination.

The new design allowed packets to travel from node to node of the network: A message from New York to Los Angeles might be routed through computers in Chicago and Denver. This meant that a new computer could be added to the network without having to connect it to every computer in the existing network. It also meant that one or more nodes of the network could be knocked out without affecting the rest of the network—a feature of packet-switching that led scientists at RAND Corporation to propose that it be used for secure digital voice communications that could survive a nuclear attack. The scientists designing ARPANET weren't worried about bombs—they were just looking for a way to share expensive computer resources. But the network's ability to survive the temporary loss of some its nodes was crucial in the early era of unreliable networking components.

Further developments in the 1970s allowed ARPANET to connect easily to other networks, and by the time the ARPANET experiment was officially ended in 1978, the idea of networked computers—along with its killer application, e-mail (see page 69)—was well entrenched.

1971 MICROPROCESSOR

The rise of the integrated circuit (see page 18) in the 1960s meant that electronics makers were cramming more and more components onto their computer chips. As the size of components continued to shrink, the next step became inevitable: putting all the components of a computer's the central processing unit (CPU), which performs basic logic operations and executes software commands—on one stand-alone chip. By the end of the decade, several companies were racing toward that goal simultaneously.

In the end, it was Intel's Ted Hoff who got credit for getting there first. Intel had been asked to develop a chip for a Japanese electronic calculator company, and Hoff and his team designed a simple generic chip packed with 2,300 transistors that could perform basic logic functions, allowing it to be programmed as a calculator. When Intel realized how widely applicable this chip could be, they bought its rights back from the calculator company. In November 1971, they released the Intel 4004 to the market with an ad that read: "Announcing a new era of integrated electronics: A microprogrammable computer on a chip."

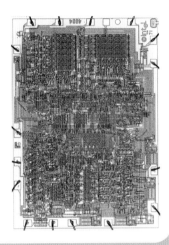

The Intel 4004 and its successors fueled the rise of the personal computer, turning Intel into an industry giant with revenues of $35 billion in 2006. But there are others who can claim to have developed microprocessors independently, most notably Gary Boone of Texas Instruments, who was granted the first patent on the microprocessor (Intel didn't file a claim, since they considered it an evolutionary development).

Intel's 4004 was a breakthrough invention that paved the way for embedding intelligence in inanimate objects as well as the personal computer.

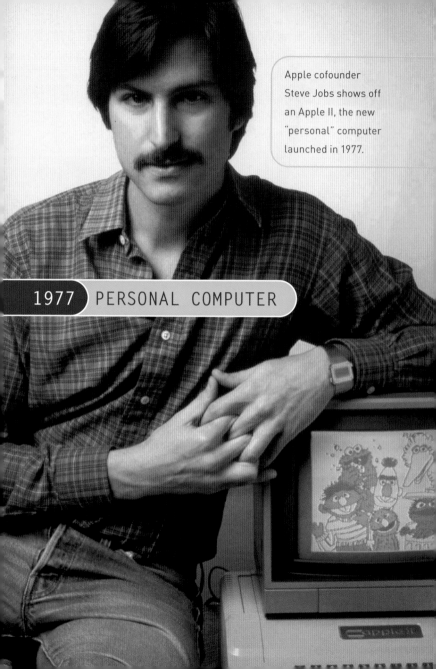

Apple cofounder Steve Jobs shows off an Apple II, the new "personal" computer launched in 1977.

1977 PERSONAL COMPUTER

"**APPLE II WILL CHANGE THE WAY YOU THINK** about computers," read a 1977 ad for a new "personal" computer—and that proved to be true. Personal computing wasn't just a matter of making computers small and cheap enough for individuals to buy. That had already been achieved in 1975 by the Altair 8800, a build-your-own computer kit selling for $397 that launched a craze among hobbyists. But the Altair computer without a keyboard or monitor, required all programs to be entered laboriously through switches on the front panel. It displayed the results with sequences of flashing lights.

Steve Jobs and Steve Wozniak, Apple's founders, realized that computers could find a wider audience if they were presented in a self-contained unit, with a keyboard, monitor, and storage device. Their first computer, the Apple, had been a crude device that they assembled by hand in Jobs' parents' garage, and they sold only about 200 of them. But the Apple II, launched in 1977 with the help of a $250,000 bank loan, was the user-friendly package they had envisioned, complete with the ability to play sounds and display text and graphics in color. By September 1980, they had sold 130,000 of them—and the computer had finally escaped from the confines of high-tech labs and big businesses, and invaded homes across the country. A few months later, the company went public, with a valuation of $1.8 billion.

> "Yes, it's okay to let your kids use Apple II."
>
> APPLE AD, 1977

Apple wasn't the only company introducing PCs. The Commodore PET was launched at around the same time as the Apple II, and Tandy Radio Shack's TRS-80 followed a few months later. All three computers were successful, though Apple received a boost when

This 1978 Apple advertisement proclaimed the arrival of "the era of the personal computer," although the term personal computer later became associated with IBM's competing product.

the first personal computing "killer app," the spreadsheet program VisiCalc, was written for the Apple II in 1979. The first what-you-see-is-what-you-get (WYSIWYG) word processor followed around the same time, followed by the first database programs.

Finally, four years after the first PCs were launched, industry heavyweight IBM entered the market with the IBM Personal Computer. The strength of the IBM brand was the final push computers needed to gain full public acceptance. The architecture chosen by IBM—using Intel 8088 chips and Microsoft software—set the basic template for PCs that has endured to the present. It guaranteed the fortunes of Intel and Microsoft, making the term "personal computer" synonymous with IBM, the latecomer to the market.

We often use the terms "Internet" and "World Wide Web" interchangeably, but the Web is actually a much more recent development, invented by Tim Berners-Lee in 1989. The Internet (see ARPANET, page 24) describes the physical network of computers and the cables that connect them, linked by a common language called Internet Protocol (IP), which was developed by Robert Kahn and Vinton Cerf in the 1970s. The World Wide Web, on the other hand, refers to the collection of documents, sound files, and videos that are "linked" to one another in an enormous decentralized web.

Berners-Lee, who was a physicist at the European Organization for Nuclear Research (CERN) in Geneva, Switzerland, when he proposed the Web, came up with three crucial components. He created "Hypertext Markup Language," or HTML, the language used to create Web pages. He also created "Hypertext Transfer Protocol," or HTTP, the method computers use to retrieve Web pages from servers. And finally, he codified the "Uniform Resource Locator," or URL, which provides a unique address (such as http://www.popularmechanics.com) for each Web page.

"Anyone who has lost track of time when using a computer knows the propensity to dream, the urge to make dreams come true, and the tendency to miss lunch. The former two probably helped."

TIM BERNERS-LEE

Berner-Lee considered several names for his new system, including Mine of Information ("Moi") and The Information Mine ("Tim"), before settling on World Wide Web. Unlike many Internet pioneers, Berners-Lee did not cash in on his contributions—but he has no regrets. The success of his World Wide Web depended on its widespread adoption, which could be encouraged only by making it freely available, he says: "You can't propose that something be a universal space and at the same time keep control of it."

Tim Berners-Lee was a researcher at CERN, the European Organization for Nuclear Research, when he proposed what eventually became the World Wide Web in 1989.

They didn't create the first World Wide Web search engine, but Larry Page and Sergey Brin came up with a search technique so powerful that the name of their company—Google—ended up in the dictionary as a verb meaning "to search on the Web." Page and Brin were computer science Ph.D. students at Stanford when they met and developed the "PageRank" system that powers their search tool.

Early search engines simply searched Web pages for the appearance of whatever key word the user searched for—the more times the key word appeared, the higher up it appeared in the list of search results. As the number of pages available on the Web mushroomed in the 1990s, this search technique became less and less effective, as good results were often buried in a sea of irrelevant or unreliable results. Page's Ph.D. research involved tracing the network of "back links" that pointed to a Web page, and figuring out which of those links came from pages with many links and which ones came from dead ends. Although it began as an academic exercise to understand the structure of the Web, Page and Brin soon realized that it provided a convenient way to rank the relevance and reliability of Web sites.

> "It was fun, it was short, it was reasonably easy to spell."
>
> LARRY PAGE on why they chose the name Google

The ranking system, which they called PageRank, after Larry, makes two basic assumptions. First, Web pages linked via many other pages are more likely to be of interest than Web pages to which no one has linked—essentially seeing the process as an

This garage in Menlo Park, California became the birthplace of Google when co-founders Larry Page and Sergey Brin agreed to rent it for $1,700 a month in 1998. The company bought the property in 2006.

Internet popularity contest. Second, not all links are of equal value—for instance, a link from a major university is more likely to indicate that a site is reliable than a link from a teenager's MySpace page. Every Web page is assigned a PageRank number between 0 and 10, which is determined by the PageRanks of the pages that link to it (each of those PageRanks being determined by the next set of links, and so on, across the whole Internet).

The result was a search engine that Page and Brin first experimented with on the Stanford campus in 1996. They launched Google in 1998, with the help of some venture capital backing. A decade later, Google's family of sites was the most-visited destination on the Web, with more than half a billion unique visitors per month—and the company was valued at more than $200 billion.

CHAPTER 02

> > > LEISURE

1951 KERNMANTLE ROPE

Rock climbers need their rope to have two key properties. First, it shouldn't break if they fall, so it needs to be strong. Second, it needs to be slightly elastic so that they're not jolted to a stop instantly if they fall. Virtually all modern climbing rope is based on a design unveiled in 1951 by the German company Edelrid called "kernmantle" rope (sometimes referred to as "core-and-sheath" rope).

Until the 1940s, most rope was made of natural fibers such as hemp or manila (a relative of hemp), woven from three strands in a helical pattern. The invention of nylon by Wallace Carothers at

Kernmantle ropes, in addition to being used in climbing and caving, are often used in sailing and other sports.

DuPont just before World War II led to new, stronger ropes made of synthetic fibers—but they still had limitations, such as the tendency to get stiff when wet, and were prone to abrasion by sharp rocks. Finally, Edelrid introduced a completely new style of rope, featuring a core made up of several long parallel bundles of twisted strands. The core was covered and protected by a woven sheath, and both parts were made of nylon.

The precise design of the new kernmantle ropes could be adjusted to produce more "dynamic" properties (extra stretch, ideal for stopping the fall of a lead climber) or more "static" properties (less stretch but more strength, better for hoisting loads or rappelling down a cliff). The improved safety of the new ropes helped usher in a new era of bolder, more adventuresome climbing.

1956 VIDEOTAPE RECORDER

In 1952, engineers at Ampex Corporation set up a demonstration of the new videotape recording system they were developing. Company executives watched in silence as a black-and-white recording of a cowboy show flashed indistinctly across the screen. "Wonderful!" Ampex's president, Alexander Poniatoff, exclaimed as the video ended. "Now which is the cowboy and which is the horse?"

Although sound recording on magnetic tape had been developed in Germany in the 1930s, storing pictures on tape presented a much greater challenge. The problem: to store the greater amount of information in a video broadcast, the tape would have to spool past the record and playback heads more than 100 times faster than in audio recorders, meaning that a half-hour length of audiotape would hold less than a minute of video—if the delicate system even survived the high-speed spooling. Still, several companies, including industry heavyweight RCA, were racing to develop a practical videotape system.

Led by Charles Ginsburg, the Ampex team implemented a system where the recording heads (which convert electrical signals into a magnetic pattern on the tape) would spin at high speed as the tape spooled past at a slower speed, allowing the data to fit on a shorter tape. Ampex unveiled its VRX-1000 system in 1956 at the National Association of Radio and Television Broadcasters convention in Chicago, surprising delegates by replaying the last few moments of the keynote speech they'd just heard—the first instant replay.

Ampex's system was a smash hit with broadcasters. At the time, broadcasting a TV show live on the East Coast and delaying it by three hours before rebroadcasting it on the West Coast required the cumbersome and expensive kinescopt process, which involved

Unlike earlier videotape recorders (VTRs), Sony's U-matic packaged the magnetic tape into a cassette, making it the first video*cassette* recorder, or VCR, when it was released in 1971.

pointing a film camera at a high-resolution TV screen, and then rushing the film out to be processed during the three-hour window. Still, those early videotape recorders (VTRs) were large, expensive, and used reel-to-reel tape. It wasn't until 1971 that Sony released its U-matic, which packaged the magnetic tape neatly in a cassette—the first videocassette recorder, or VCR.

1958 | POLYURETHANE "FOAM" SURFBOARD

In December 2005, Gordon "Grubby" Clark ordered his factory to stop producing the foam "blanks" that until then had been shaped into surfboards, and he instructed his workers to smash the molds and cut up the specialized machinery with torches. Given that Clark Foam supplied some 90 percent of the world's foam blanks, the news hit surfers hard, and prices doubled almost overnight as surfboard makers scrambled to find alternative suppliers. The puzzling shutdown, attributed in part to Clark's difficulties in meeting ever-stricter environmental regulations, highlighted an important fact: nobody could make surfboard blanks quite like the man who, along with his mentor Hobie Alter, had invented the modern foam-and-fiberglass board almost half a century earlier.

The history of the surfboard dates back to at least A.D. 400, when Hawaiian islanders caught waves on long redwood planks. Since the sport reemerged at the beginning of the twentieth century,

"And I thought, by God, this is really it."

HOBIE ALTER, after experimenting with the first polyurethane surfboards the board has been constantly evolving—getting shorter and lighter, as designers experimented with single or double fins and subtle shape changes. In the 1950s, the standard surfboard was made of lightweight balsa wood encased in a fiberglass skin. These boards were a vast improvement over previous designs, but balsa wood was imported and hard to come by, and it absorbed water if the fiberglass was damaged. Various surfboard builders had been experimenting with plastics like Styrofoam since the 1940s, but none had managed to replace balsa.

Alter had opened Hobie Surfboards in 1954, and in 1957 he hired Clark to help him make boards. Once the pair began to experiment with polyurethane, a rigid, lightweight foam, they

Surfing was revolutionized in the 1950s by the development of fiberglass-coated boards with a core of polyurethane foam.

knew it had the potential to revolutionize surfboard design. It took months of experimentation to get the right chemical mix, with Alter driving loads of deformed or porous blanks to the San Juan dump, but they finally sold their first polyurethane board, coated in fiberglass, in 1958. By 1961, the business had expanded so much that Clark branched off to start Clark Foam, supplying just the blanks that board makers like Alter would then shape into surfboards. With Clark now out of the picture, other manufacturers are rushing to fill the hole in the $200-million-a-year market—or to develop new materials for the surfboards of the future.

1962 VIDEO GAMES

The PDP-1 computer that arrived on MIT's campus in 1961 was a serious, $120,000 research tool. It was the first of a new generation of computers that was smaller and easier to manipulate than its predecessors (see minicomputer, page 20). "And it was just sitting there," a programmer named Steve Russell later recalled. To put the new computer to work and explore the its capabilities, Russell and a group of fellow programmers wrote the first interactive computer game, which they dubbed "Spacewar." Inspired by the science fiction novels of E. E. "Doc" Smith, Spacewar featured two spaceships that could fire missiles at each other and launch into hyperspace while avoiding the gravitational pull of the Sun. The programmers made the code freely available, and the game quickly spread to campuses across the country—where other students also had free access to the hugely expensive computers of the time.

Dan Edwards (left) and Peter Samson, shown playing "Spacewar" in 1962, were two of the MIT computer enthusiasts who wrote the first video game for the university's PDP-1.

One of those campuses was the University of Utah, where an engineering student named Nolan Bushnell played the game in the mid-1960s. In 1971, Bushnell put together his own version of the game with a stand-alone machine, jury-rigged with a coin slot from a laundromat, to create the first arcade game, called Computer Space. The game flopped, but Bushnell and his fledgling company Atari followed up a year later with the smash hit Pong, launching the video-game era.

"When you divide 25 cents into an $8 million computer, there ain't no way."

Nolan Bushnell, on why cheap arcade consoles had to be developed before the first video games could become popular

MUSIC SYNTHESIZER

Robert Moog, the founder of the pioneering company Moog Music, did not invent the electronic synthesizer. By the time he introduced his "voltage controlled electronic music module" in 1964, other music-synthesizing machines had been around for more than a decade. But Moog's was the first to make the leap from science experiment to musical instrument. The secret to its success was that it had higher-quality sound and was controlled with a piano-style keyboard rather than by punch cards—an innovation suggested by Moog's collaborator, the composer Herb Deutsch.

Moog had responded to the needs of musicians, and they in turn were quick to adopt his machine: the Monkees used one in 1966, the Rolling Stones in 1967, and the Beatles in 1969. The turning point for the general public was the 1968 release of Walter Carlos's innovative and exciting *Switched-On Bach*, a collection of Bach pieces carefully recorded one line at a time, which earned

This prototype Moog synthesizer, built in 1964, was heard in public for the first time at New York City's Town Hall on September 25, 1965.

three Grammy Awards and became the first classical album to earn platinum status by selling half a million copies.

The original Moog was still too unwieldy for most live performances, leading Moog to follow up with the Minimoog in 1970, which could be folded up and carried like a suitcase. Other manufacturers were quick to follow suit, and electronically synthesized sounds soon became mainstream. But Moog's influence wasn't just aesthetic: the widespread acceptance of electronic music paved the way for the audio technology underlying today's computer sound cards, cell phones, and stereos.

> "I'm a toolmaker. I design things that other people want to use."
>
> ROBERT MOOG

1970 DIGITAL MUSIC

In 1982, Billy Joel's *52nd Street* became the first album to be released on compact disc, following the unveiling of the new technology by codevelopers Philips and Sony. By 2000, more than 2.5 billion CDs were being sold each year—and though that number is now declining, the downloadable music that is replacing it also relies on the fundamental innovation of storing music digitally as a binary series of 0s and 1s. It was a major coup for the two electronics giants, but the original idea was actually conceived—and patented—by a lone inventor back in 1970.

James Russell was a scientist at Pacific Northwest Laboratory in Richland, Washington, and an avid music fan. Frustrated by the way his vinyl records deteriorated after repeated playing (even when he tried tricks such as using a cactus needle as the record-player stylus), he was looking for a way to play music that wouldn't require physical contact between the moving parts. Using a laser for a digital readout would work, he realized—if he could somehow encode the music as alternating patches of dark and light.

It was already known that the sound of a voice could be transformed into a stream of 0s and 1s: during World War II, the United States and its allies had developed a secure phone line that encoded conversations between their leaders in a digital format. But the idea of storing music that way met resistance from Russell's bosses when he proposed it in the late 1960s. "'Music into numbers? Come on now, Russell'" was their response, he later recalled. He patented his digital-to-optical recording system in 1970 but met the same response from major electronics companies. They later

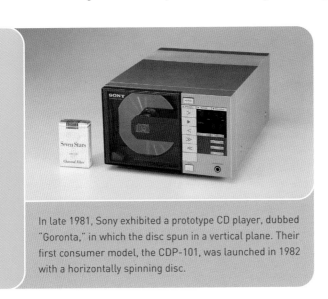

In late 1981, Sony exhibited a prototype CD player, dubbed "Goronta," in which the disc spun in a vertical plane. Their first consumer model, the CDP-101, was launched in 1982 with a horizontally spinning disc.

regretted it, though: in 1992, Time Warner and other CD manufacturers paid a $30 million patent infringement settlement to Russell's former employer.

As technology continued to improve in the 1970s, the benefits of digital music eventually became clear, leading to the Sony–Philips collaboration. Although Russell had developed the basic idea, plenty of details had yet to be worked out—such as the size of the disc. Philips wanted it to be small enough to fit in the jacket pocket of a suit, but Sony executive Norio Ohga insisted on a size of 12 centimeters (just under five inches), which would hold 75 minutes of music: "Just as a curtain is never lowered halfway through an opera," he said, "a disc should be large enough to hold all of Beethoven's Ninth Symphony."

> "I use the analogy of the Wright Brothers' development of the airplane. They weren't successful, but they showed the whole world you could do it, in principle."
>
> JAMES RUSSELL, on his initial failure to convince others to adopt his CD system

1971 KEY-FRAME COMPUTER ANIMATION

The 1995 smash hit *Toy Story* was the first full-length computer-animated feature film. Its startlingly realistic three-dimensional action helped launch a new movie genre—and it drew attention to some of the long-forgotten pioneers of computer animation. A year later, Nestor Burtnyk and Marceli Wein received an Academy Award for technical achievement, recognizing their development of "key-frame" computer animation in the late 1960s and early 1970s.

The key-frame method of animation was established long before computers came on the scene. A senior artist would draw the most important, or "key," moments in any sequence, and then junior artists would perform the tedious task of drawing all the intermediate frames necessary to give the illusion of smooth movement. Burtnyk and Wein were researching human-computer interaction at Canada's National Research Council, experimenting with ways to make it easier for artists to interact with the unwieldy computers of the era. They realized that computers would be perfectly suited to perform the mindless task of filling in the intermediate steps between key frames.

Working with an artist named Peter Foldes, they developed a system wherein the animator could draw key frames using a light pen, then manipulate his pictures using a mouse (a little-known

The vividly animated three-dimensional characters in the 1995 hit movie *Toy Story* relied on computer animation techniques pioneered more than two decades earlier.

gadget at the time). The team presented their first experimental short film using the new technique in 1971, and they followed with *Hunger* in 1973, which became the first computer-animated film to be nominated for an Oscar, in the category for best short film.

1972 WAFFLE-SOLE RUNNING SHOES

Bill Bowerman, the track coach at the University of Oregon, knew his athletes would run faster if they had lighter shoes. So he built his own, pouring urethane rubber into his wife's waffle iron to make the soles after a breakfast-time moment of inspiration in 1971. The first prototype waffle-sole shoes (dubbed "Moon Shoes" because the distinctive tread resembled Neil Armstrong's footprints on the Moon) were worn by runners in the 1972 Olympic Trials. Two years later, the Waffle Trainer was released by the company Bowerman had founded along with one of his former Oregon athletes, Phil Knight. The shoe was an instant hit for Nike, as the young company was known.

> "You can bet there wouldn't be so many people running today if they had to carry all the extra baggage we had back then."
>
> BILL BOWERMAN

It wasn't just Olympic runners who liked the new shoe. During a trip to New Zealand in the 1960s, Bowerman had been surprised to see how many people of all ages and sizes were out running for fun and fitness. On his return to Oregon, he coauthored a guide called *Jogging*, which sold over a million copies and helped spark a running craze in the

The first athletes to try out Bill Bowerman's new waffle-sole shoes were members of his track team at the University of Oregon, such as U.S. record holder Steve Prefontaine, seen here racing to fourth place in the 5,000 meters at the 1972 Olympics in Munich.

United States. That movement has never really subsided—thanks in large part to Bowerman's creation of lightweight shoes that are as comfortable for jogging as they are for everyday wear.

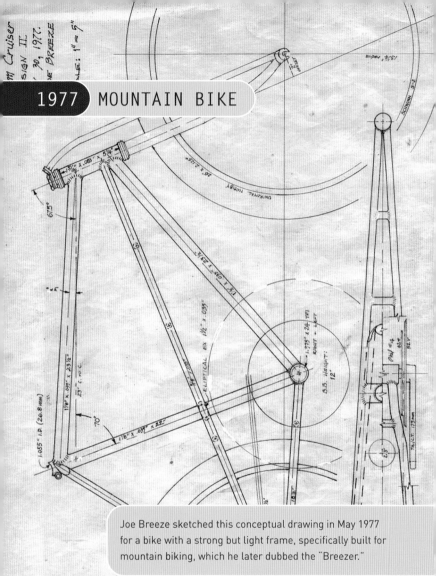

Joe Breeze sketched this conceptual drawing in May 1977 for a bike with a strong but light frame, specifically built for mountain biking, which he later dubbed the "Breezer."

IT'S HARD TO PINPOINT who first put fat tires and flat handlebars on a bike to help it handle rough terrain—experiments like that have been going on for at least a century. But the modern mountain bike emerged from a group of riders and bike makers in Marin County, just north of San Francisco, who in the 1970s began to gather and race down a fire road on Mount Tamalpais that drops 1,300 feet in two

"It was a big hot secret, like a new drug. You'd turn someone on to it, and they were like, 'Wow, I've gotta have some of that. I gotta have one of those.'"

GARY FISHER, on the first mountain bikes

The first "Breezer," built from scratch in 1977, now resides in the Oakland Museum's Cowell Hall of California History.

miles. In their quest to go faster down the hill—and to be able to ride back up to do it again—some of the riders began building more and more innovative bikes.

One of the pioneers was Joe Breeze, who welded together the first custom-made frame designed specifically for mountain biking—strong but light—in 1977. Breeze built 10 of these early bikes, dubbing them "Breezers." In 1979, another Marin County rider, Gary Fisher, formed a company called Mountain Bikes with several partners to begin selling their new designs. "We were the only place where you could walk in, pay your money, and ride out," Fisher later recalled.

As the new pastime swept the country, designs continued to evolve toward lighter, stronger, and more stable two-wheeled vehicles, including those made of titanium alloy. One of the biggest steps came in 1988, when Kestrel unveiled a carbon-fiber bike designed by Keith Bontrager featuring a motorcycle-inspired suspension fork designed by Paul Turner. By the turn of the century, an estimated 20 million people had caught the mountain biking bug.

1999) DIGITAL VIDEO RECORDER

When the first digital video recorders appeared in stores in 1999, most people didn't immediately recognize them as revolutionary. After all, nearly everybody already had a box that could record and replay TV shows. "You describe it, and everyone goes,

Digital video recorders like this 1999 TiVo model changed how people watch TV—and threatened the advertising model on which broadcasters rely.

'Oh, that's just like a VCR,'" said Mike Ramsay, cofounder of the pioneering DVR company TiVo. "That's just like telling Hindenburg it was just a balloon."

What TiVo and other competing products did was to record TV broadcasts on a computer hard disk, making it easy to organize and instantly access programming with no fast-forwarding or rewinding of tapes. But its most revolutionary feature was that it could be used "live": viewers could rewind while watching live television to catch something they missed—and even watch it in slow motion—and then continue watching right where they left off. Or they could begin watching a show 15 minutes after the broadcast started, and fast-forward through any parts they didn't want to watch until they caught up to real time. It's this last part—the ability to skip commercials—that proved to be the major selling point for DVRs and made them the target of fierce resistance from the television industry.

TiVo was started in 1997 by Ramsay and Jim Barton, Silicon Valley engineers who had worked previously at Silicon Graphics on "interactive television," an early 1990s idea that was superseded by the rise of the Internet. They faced competition initially from a company called Replay Networks, which had devised a similar product, and later from generic DVRs marketed by cable companies directly to their customers. By 2007, more than 18 million households had DVRs, including 4.4 million TiVo subscribers. Whether or not TiVo itself manages to survive the fierce competition in the market it created, Ramsay and Barton's legacy— uniting the TV and the computer—is undoubtedly reshaping the nature of home entertainment.

CHAPTER 03

>> COMMUNICATION

ANSWERING MACHINE

The earliest answering machine design goes back—like so many other things—to Thomas Edison, who suggested a telephone recording system in 1877, less than a year after Alexander Graham Bell's first demonstration of the telephone. Edison's machine didn't work very well, and he soon abandoned it—but dozens of other inventors followed in his footsteps over the succeeding decades, patenting ideas and building prototype answering machines. Nothing managed to catch on—partly because AT&T was vehemently opposed to the technology, attempting to preserve its monopoly on phone equipment by refusing to allow non–AT&T devices to be connected to its phone lines. This policy was backed up by Federal Communications Commission regulations, effectively preventing any home answering machine from catching on.

In 1948, an engineer in Wisconsin named Joseph Zimmermann devised a new answering machine—to get around the AT&T prohibition, his machine didn't need to be directly connected to the phone line. When the phone rang, a lever arm mechanically lifted the phone's receiver off the hook. A 45-rpm record then played the outgoing message, with the sound traveling through the air to the receiver rather than through wires connected to the phone, and the incoming message was recorded on a strand of steel wire. In 1949, Zimmermann cofounded a company called Electronic Secretary to produce his answering machine, which was so successful that it helped force the FCC to loosen the rules governing answering machines.

Still, by 1957, AT&T estimated that only 40,000 customers were using answering machines. As electronics improved and

Early answering machines like this Ansafone Mk VII from about 1970 were slow to catch on until the regulations governing home telephone equipment were loosened in the 1980s.

magnetic recording was adopted, answering machines became smaller and more convenient. But it wasn't until the AT&T phone monopoly was broken up in 1984, loosening the rules around consumer telephone equipment, that answering machines really took off.

1948 RADIO-FREQUENCY IDENTIFICATION

Radio-frequency identification, or RFID, is an idea with enormous potential that was understood long before the technology to make it practical existed. If you want to identify an item—a cow, or an airplane, or a box of eggs—you tag it with an RFID transponder that consists of a small antenna and a microchip. You can then remotely identify the item by sending radio waves that are picked up by the transponder, which sends a signal back containing the information encoded on the microchip (a serial number, for example).

One of the earliest uses of this basic technique was during World War II, when British planes were equipped with transponders that picked up British radar signals and sent back a distinctive signal to identify them as friendly planes. In 1948, a U.S. Air Force engineer named Harry Stockman wrote a seminal paper titled "Communication by Means of Reflected Power," laying out the basic principles of an RFID system that would transmit more complex information. For the next 30 years, it was simply a matter of technology catching up with Stockman's vision.

In the 1960s, "electronic article surveillance" tags were introduced to prevent theft from stores. These were simple RFID tags that broadcast just one bit of information: The cashier would switch the bit from "0" to "1" when the item was paid for, allowing the customer to exit without triggering an alarm. In the 1970s, a host of more complex applications were developed: remote toll collection systems for highways, implantable tags to help keep track of which cows had received their medication, access cards to unlock doors. Now, as miniaturization allows RFID tags to be smaller than a grain of rice and cost only a few cents each, large retailers like Wal-Mart are moving to implement RFID tracking throughout their supply chain in a bid to improve inventory control. Security at ports will also be improved

> "Evidently, considerable research and development work has to be done before the remaining basic problems in reflected-power communication are solved, and before the field of useful applications is explored."
>
> HARRY STOCKMAN in 1948, correctly predicting that his RFID scheme was not yet practicable

The basic principles of radio-frequency identification were laid out in a 1948 paper, but it took several decades to develop small and inexpensive chips like this RFID sticker tag, which incorporates a flat antenna spiraling around the center.

when the contents of all shipping containers can be read instantly and remotely.

RFID boosters look forward to the day when everyday objects are tagged, creating an "Internet of things" that communicate with each other. Clothes could tell the washing machine what setting to use, for example. Critics point to privacy concerns, but the technology isn't ready anyway—for now.

1949 BAR CODE

The lines of the first bar code were traced in the sand of Miami's South Beach in 1949. Philadelphia engineer Joe Woodland was on vacation from Drexel Institute of Technology, where he and

The Universal Product Code, developed by IBM in the 1970s, created a standardized format for bar codes to be read out by a "wraparound" web of laser light.

colleague Bernard Silver had been pondering ways to automate the supermarket checkout process. As he absentmindedly drew his fingers across the sand, he realized that the lines could represent the dots and dashes of Morse code, perfect for a product code that could be read using ultraviolet light. He and his colleague patented the idea in 1952—and then waited.

It wasn't until a decade later, with the invention of the laser (see page 204), which provided a better readout, and the continued shrinking of computers, that interest in the bar code picked up. The railway industry experimented in the 1960s with bar codes on railcars to monitor their whereabouts, and a Kroger's supermarket in Cincinnati ran a bar code trial in 1972. But widespread adoption required both supermarkets and suppliers to agree on a standard format for bar codes, an achievement that was finally realized in 1973 when a proposal by IBM's George Laurer, building on Woodland's work, was accepted. The code contained 12 digits, including one "error-check" digit that was calculated as a mathematical function of the first 11 numbers to ensure that the code was read accurately. (New bar codes have 13 digits to allow more information to be encoded.)

Finally, in June 1974, a pack of Wrigley's Juicy Fruit gum became the first product scanned with a Universal Product Code (UPC), at a Marsh Supermarket in Troy, Ohio. The system wasn't an instant hit: *Business Week* ran a 1976 story headlined "The Supermarket Scanner That Failed," noting that only about 50 stores, rather than the anticipated 1,000, had installed scanners. But the benefits gradually became obvious: the scanners yielded not just faster checkouts, but better inventory tracking, and detailed information about customer purchasing habits. Now, bar codes are used at every stage of the manufacturing and supply chain process, and about five billion bar codes are scanned every day.

1960 MAGNETIC STRIPE CARD

The United States reached a tipping point in 2003: for the first time, credit and debit cards were used more often than cash and checks to pay for in-store purchases. The technology that made this possible was a simple magnetic stripe that encodes such information as name, account number, and expiration date—an unexpected by-product of IBM's pioneering research in magnetic recording for computer storage in the 1950s.

IBM researchers were trying to develop an identity card for government workers that was secure against counterfeiting when they considered attaching a piece of magnetic tape to the card to store information "invisibly." The problem was figuring out how to

Tickets with magnetic stripes made their debut at Chicago's O'Hare International Airport in 1970, thanks to a joint IBM-American Airlines project.

affix the delicate tape to the card without destroying it. An IBM engineer named Forrest Parry finally solved the problem, using his wife's iron to develop the "hot-stamping" process, and he published his results in 1960.

It took another decade before the process reached American consumers. In 1970, American Airlines and IBM introduced ticketing with magnetic stripes at Chicago's O'Hare International Airport. The Bay Area Rapid Transit system in San Francisco followed suit with magnetic stripe fare cards in 1972—a plan that hit a snag when a local scientist demonstrated that he could copy fare cards and change the monetary value encoded on the stripes. IBM went back to the drawing board for a year and produced a tamper-proof stripe—much to the relief of banks and credit card companies, who were in the midst of devising industry-wide standards for plastic cards with magnetic stripes.

The new cards allowed the establishment of the first electronic authorization system for credit cards in 1973, sparing merchants the necessity of calling the credit card company before each purchase. The result: Visa alone now processes some 74 billion authorizations each year, and its automated system can handle 12,000 transactions per second.

Some ideas are so revolutionary that they seem like fantasy when they're proposed—so it's not surprising that it was Sir Arthur C. Clarke, best known as the author of *2001: A Space Odyssey*, who first wrote about satellites that "could give television and microwave coverage to the entire planet," in a 1945 article in *Wireless World*. Seven years later, an engineer at AT&T's Bell Laboratories named John Pierce came up with the same idea independently. But the idea was still so far-fetched that he published it as a story in *Astounding Science Fiction* under the pseudonym J. J. Coupling.

When Sputnik (see page 201) was launched in 1957, becoming the first artificial satellite, Pierce saw new possibilities for his idea. "For me, space had suddenly changed from science fiction to a technology that could be put to some sensible use," he wrote later.

About 700 scientists and technicians from Bell Labs worked to prepare the first communications satellite, Telstar I, for launch in 1962.

At Bell Labs, he guided a project to send a 100-foot-diameter Mylar balloon coated with aluminum into space. The "Echo" mission bounced microwave signals off the reflective balloon, sending a recording of President Dwight D. Eisenhower's voice from California to New Jersey in 1960.

Echo's "passive" signal-bouncing capability was not sufficiently advanced for truly global communications, so the next step was to build an "active" satellite that would amplify and retransmit the microwave signals. About 700 Bell Labs scientists and technicians tackled the problem, and in 1962 Telstar I was launched, becoming the first modern communications satellite. Vice President Lyndon B. Johnson, in Washington, was the recipient of the first long-distance telephone call relayed by satellite, from a transmitting station in Andover, Maine; later the same day, the first live transatlantic television images were sent to a receiving station in France.

The network of satellites that followed Telstar I quickly became indispensable—as millions of Americans discovered when a single communications satellite, Galaxy IV, malfunctioned in 1998. About 30 million pagers went silent, ATM and credit card payments were disrupted, and some television stations simply stopped broadcasting because of a lack of available programming.

"It's like a writer of detective stories going home and finding a body in his living room."

JOHN PIERCE, engineer and science fiction author, on his reaction to the Sputnik launch

1963 | TOUCH-TONE DIALING

WHEN AT&T INTRODUCED TOUCH-TONE DIALING in 1963, they pitched it to their customers as a time-saving device, claiming that it reduced average dialing time for a 10-digit number to four seconds, compared with 14 seconds for old rotary-dial phones. As the testimonial from an unnamed customer in a 1964 press release put it, "It cuts dialing time in half, and to a businessman like myself, this means saving money."

In fact, the new dialing system had a number of benefits that went beyond saving a few seconds. Each button produces a tone made up of two possible frequencies: the rows on the keypad correspond to different low frequencies, and the columns to different high frequencies. These signals could be interpreted by the switching equipment and were more easily transmitted across long distances and poor-quality lines than rotary-dial signals, paving the way for direct long-distance dialing. The system's designers were also looking ahead to the possibility of interacting directly with computers via the tones, leading to today's much-maligned automated call systems.

"The whole thing is like magic. You can dial very fast and it's just wonderful."

AT&T CUSTOMER testimonial, 1964

In addition to the electrical engineering needed to make the new system work, AT&T invested considerable effort in the "human factors" engineering when designing a suitable keyboard. They considered 16 different keyboard layouts, ranging from grids and circles to crosses and diagonal lines, and ran head-to-head tests to see which ones were preferred. The modern telephone keypad, reading 123-456-789-0 (left to right and top to bottom), emerged victorious—although it ran opposite to the more common adding-machine layout that is still used for calculators and computers (789-456-123-0).

A Bell Labs technical report from 1960 shows that the adding-machine layout was eliminated early in the trials by a horseshoe-shaped "speedometer" layout, although its average keying time of 5.08 seconds was hardly different from the 4.92 seconds of the eventual winner. Enter the conspiracy theory: while Touch-Tone keypads ushered in a new level of speed in dialing, many central telephone offices still relied on antiquated electromechanical switches. By some accounts, Bell scientists feared that accountants with vast training on adding machines would be able to dial too fast for these older switches, so they eliminated that configuration.

1970 FIBER OPTICS

Optical fiber is the invention that allows us to see around corners. Light typically travels in a straight line, but if you shine it into one end of an optical fiber—basically a long, flexible string made out of glass—it will shine out the other end, even if the cable is curved or looped. The secret is internal reflection: the light bounces back and forth along the inner walls as it travels along the cable.

The term "fiber optics" emerged in the 1950s, when researchers first realized that light could travel much farther without leaking away if glass fibers were "clad" with a surrounding layer of glass with slightly different reflective properties. Although clad optical fiber improved light transmission enough to allow effective endoscopes for minimally invasive "keyhole" surgeries to be developed in the

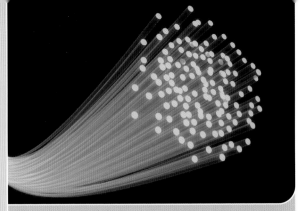

Light travels through optical fiber like water through a (slightly leaky) pipe, allowing it to bend around corners and fit through small openings.

1960s, the light could still travel only a few feet before impurities in the glass scattered the rays.

With the invention of the laser (see page 204), engineers realized that using optical fibers to send information would be vastly more efficient than the copper wires used for long-distance transmission at the time—if they could develop fibers that would successfully transmit a signal for 1,000 yards and have just 1 percent of the initial light make it to the other end.

A four-year quest to produce such a fiber ended in 1970 when Donald Keck, Bob Maurer, and Peter Schultz of

"History has clearly demonstrated that as long as scientists and engineers around the globe continue to invent the technology to transmit information at a lower cost per bit, creative people will find a reason to send proportionally more bits."

DONALD KECK, coinventor of telecommunications-grade optical fiber

Corning reached that milestone. Unlike their competitors, who were struggling to eliminate impurities from ordinary glass, the Corning team started with ultrapure fused silica and figured out how to stretch it into fibers. The result: a single modern optical fiber, slightly thicker than a human hair, can transmit more than 60 million telephone calls simultaneously, and a single pound of optical fiber can replace more than two tons of copper cable.

1971 ELECTRONIC MAIL

Try as he might, Ray Tomlinson simply doesn't remember the contents of the first e-mail. "The test messages were entirely forgettable and I have, therefore, forgotten them," he says. It was 1971, and Tomlinson was a researcher with BBN Technologies, which had been contracted to help develop the precursor of the Internet (see ARPANET, page 24). The concept of sending "electronic" messages back and forth between or among multiple users of a single large computer had existed for at least a decade. But it was Tomlinson who decided to merge the functions of two programs: one that sent mail to other users of the same computer, and another that allowed data to be transferred from one computer to another via the fledgling ARPANET network. The result: a program that could send e-mail across the network.

Tomlinson's first test messages were—as far as he can recall—gibberish sent from one PDP-10 computer to another one alongside it. Then he sent the first group e-mail: a message to other users explaining the new program. The key that users had to remember: to send a message to someone on another computer, separate the

These two computers were sitting right next to each other in 1971, but they were connected only via the ARPANET network, allowing Ray Tomlinson to test his new e-mail program by sending a message from one to the other.

username from the name of the computer with the "@" symbol. Tomlinson chose the now-iconic symbol for two reasons: it was a character unlikely to appear in anyone's username, and it signified the word "at."

No one, including Tomlinson, expected e-mail to become as popular as it did. Shortly after the first demonstration of the program, one of Tomlinson's colleagues recalls him saying, "Don't tell anyone! This isn't what we're supposed to be working on." But e-mail caught on very quickly, and rapidly became the biggest source of traffic

"I am frequently asked why I chose the at sign, but the at sign just makes sense."

RAY TOMLINSON
on the origins of "@" in e-mail addresses

fueling the growth of the Internet. Now an estimated two million e-mails are sent around the world every second.

A historical footnote: In 1978, a marketer named Gary Thuerk was planning a trip to the West Coast to demonstrate Digital Equipment Corporation's new line of computers. To drum up interest, he had an assistant type in the addresses of all 600 people on the West Coast with e-mail accounts (there was a printed directory at the time). "We invite you to come see the 2020 and hear about the DECSystem-20 family," he wrote. It was the first spam message—a category that now accounts for up to 90 percent of e-mail messages.

1973 CELL PHONE

The first call from a modern cell phone was made in April 1973, from the corner of 56th Street and Lexington Avenue in Manhattan. Using a foot-long two-pound prototype of what later became the DynaTAC, Motorola researcher Martin Cooper called Joel Engel, the head of cell phone research at rival AT&T's Bell Labs. "Joel," he said, "I'm calling you from a *real* cellular phone."

The idea for a cellular phone system—dividing cities into small hexagonal "cells" with a transmission tower in each one, so that calls could be passed on from tower to tower as the user traveled through different cells—had been proposed by Bell Labs researcher Douglas Ring in 1947, but the necessary technology didn't yet exist. Instead, mobile telephone service through the 1950s and 1960s was restricted to radio-style car phones with severe limitations. In New York City, for instance, the system could accommodate only 2,000 subscribers, and only 12 calls could be made at a time, leading to long waits.

Motorola's DynaTAC phone, weighing in at a svelte 28 ounces, became the first commercially available cell phone when it was approved by the FCC in 1983.

It was Engel and his colleagues at Bell Labs who, in the 1960s, revived the idea of a cellular network. The Federal Communications Commission was considering reallocating part of the frequency spectrum—AT&T proposed a modern cellular system that would serve an expanded network of car phones, and it asked the FCC for a monopoly over the new frequencies. Motorola, however, also wanted to enter the business, and to convince the FCC that other companies should be allowed to use the frequencies, it launched an effort to develop a handheld phone that could be used without a car. Thanks to the continued miniaturization of integrated circuits (see page 18), Cooper was able to shrink the 30-pound car units in use at the time to the two-pound prototype that made the first call.

The Motorola demonstration helped convince the FCC not to award AT&T a monopoly, but the wrangling over the frequency spectrum continued for another decade. Ignoring the advice of its marketing people, who had concluded that there was no market for cellular service at any price, AT&T ran the first large trial of the new system

"We were surprised, because we thought New Yorkers were so blasé."

MARTIN COOPER of Motorola on the stir created when he made the first cell phone call from a New York City street corner in Chicago in 1978. Finally, in 1983, commercial cellular service became available to the public.

1999 MOBILE E-MAIL

The tech gadget that defined the end of the millennium was the BlackBerry, a wireless e-mail tool developed by a little-known Canadian company called Research in Motion (RIM) and released in 1999. RIM's researchers were not the first to experiment with wireless e-mail, but they were the ones who made it work, earning the company a head start of several years on its competitors and a customer base that reached eight million in 2007. One of the keys to its success was the distinctive design of its handset, which featured a full alphabetic keyboard laid out so that two-thumb typing could be fast and convenient.

But the handset (named because its miniature buttons looked like the drupelets of a blackberry) is only the most visible part of a vast system that relies on a BlackBerry Enterprise Server, working in conjunction with a corporate e-mail server, to implement a "push e-mail" service. Most competing mobile e-mail services are "pull" systems, requiring the user to connect periodically to the server and download any new e-mail. BlackBerry's "push" system, on the other hand, sends new e-mail from the server to the handset as soon as it arrives, alerting the user in real time.

Who should get credit for the underlying technology is a matter of considerable dispute. RIM was founded in 1984 by University of Waterloo student Mike Lazaridis, who began focusing the company's efforts on wireless data in 1987. Between 1989 and 1991,

Research In Motion introduced the "RIM Inter@active Pager 950" in 1998 with a suggested retail price of $359. The device's resemblance to a blackberry prompted a name change for later versions.

his team experimented with wireless e-mail in collaboration with Ericsson and AT&T. In 2006, RIM settled a lengthy patent dispute by paying $612.5 million to a company called NTP, whose assets consist almost exclusively of patents based on the idea of wireless e-mail, which it has never implemented.

After the settlement, an unexpected postscript emerged. In 1982, a Silicon Valley computer hobbyist named Geoff Goodfellow had conceived of the idea of wireless e-mail, and he had written a brief paper describing it. He later formed a company, and in 1991 he introduced a wireless e-mail service called RadioMail, which failed to take off—but whose existence could have posed a threat to NTP's claims. During its court case against RIM, NTP hired Goodfellow as a consultant, paying him $19,600 for several days' work, along with an agreement not to publicly discuss his work until the case was settled.

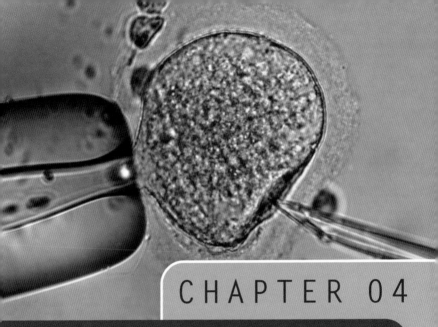

CHAPTER 04

CARBON DATING

Scientists trying to unravel the mystery of when humans first arrived in the Americas have uncovered a number of tantalizing clues, such as a pair of 13,000-year-old thigh bones found on an island off California, and firepits in Southern Chile containing 14,000-year-old debris. We know this thanks to a technique devised in 1947 by Willard Libby, then a physical chemist at the University of Chicago. Libby had worked on the Manhattan Project to develop the atomic bomb during World War II, and his experience with radioactive isotopes helped him to revolutionize archaeology.

Most carbon atoms contain either six or seven neutrons. But the cosmic rays that constantly bombard the Earth produce trace amounts of carbon-14 (sometimes called radiocarbon), a radioactive isotope that contains eight neutrons. These carbon-14 atoms slowly "disintegrate" back to normal carbon over time, with an average half-life of just over 5,700 years. As long as an organism is alive, the level of carbon-14 stays roughly constant. As Libby later explained, "All the radiocarbon atoms that disintegrate in our bodies are replaced by the carbon-14 contained in the food we eat." But once an organism dies, the amount of radioactive carbon decreases steadily over thousands of years, providing a simple but reliable archaeological "clock."

> "Now you'll turn us all into damn chemists."
>
> Noted archaeologist
> W. DUNCAN STRONG
> in 1948, after attending
> a lecture by
> Willard Libby

To demonstrate the technique, Libby and his collaborators had to devise a method for detecting the extremely small amount of carbon-14 in their samples, as

Willard Libby, shown here in 1968, earned the 1960 Nobel Prize in Chemistry for his role in the development of carbon dating.

well as show that "living" and "dead" carbon have different characteristics. They accomplished this demonstration by comparing the carbon-14 signature of methane from oil deposits with methane from a Baltimore sewage plant. Ultimately, they were able to construct a "Curve of Knowns," comparing their carbon dates with known artifacts, such as a California redwood felled in 1874, with more than 2,900 rings in its stump, and a bread roll that was overcooked when Pompeii was buried under lava in A.D. 79. The results were remarkably accurate, and with some modifications, scientists can now accurately date organic artifacts dating back 50,000 years or more.

HIGH-YIELD RICE

In 1968, the administrator of the humanitarian aid agency USAID, William Gaud, coined the term "Green Revolution" to describe the massive changes in agriculture that doubled world grain production between 1950 and 1976. Although the widespread use of fertilizers, pesticides, and irrigation played a role in this revolution, the most significant changes came from new varieties of rice and wheat

The "miracle rice" IR-8 had a dramatic impact on crop yields when it was introduced in 1966, though it produced a "chalky" grain that shows up as opaque rather than translucent in this image. Later dwarf varieties produced a tastier, less chalky grain without sacrificing yield.

developed by scientists. These new varieties were chosen to grow quickly, resist diseases, and produce as much nutrition as possible. In early domesticated food crops, about 20 percent of the energy harvested from the Sun through photosynthesis went to producing edible seeds; in contrast, in modern wheat, rice, and corn, more than half of that energy goes to seed growth (while saving energy by, for example, growing shorter stalks in "dwarf" varieties).

About half the world's population relies on rice for daily sustenance. In 1960, the Ford Foundation and Rockefeller Foundation set up the International Rice Research Institute in the Philippines with the goal of improving rice yields, especially in the developing world. The breakthrough came in 1966 with the release of IR-8, a cross between a tall Indonesian variety called Peta and a semidwarf from Taiwan called Dee-Geo-Woo-Gen. Dubbed "miracle rice" by the press when it was released, IR-8 had an immediate impact: by 1968, the Philippines was self-sufficient in rice production for the first time in decades. The new rice quickly spread throughout Asia, and further breakthroughs followed, such as the development of the first hybrid rice by Yuan Longping in China in 1974.

The rice research had followed earlier work on wheat, begun in the 1940s by Norman Borlaug and his colleagues at a Rockefeller-funded research institute in Mexico. By the 1960s, Borlaug's high-yielding, short-stalked, disease-resistant wheat strains had

> "There is no longer any excuse for human starvation."
>
> GEORGE HARRAR, the Rockefeller Foundation scientist who established the International Rice Research Institute, on the development of the IR-8 rice strain

proved themselves viable around the world, and he was awarded the Nobel Peace Prize in 1970. While the initial goal of the Green Revolution—eliminating hunger—remains elusive, the development of high-yield grains has undoubtedly transformed modern agriculture.

1975　SUCRALOSE

Inventions sometimes are stumbled upon by accident—with artificial sweeteners, in fact, they always seem to happen by accident. The first artificial sweetener, saccharin, was discovered in 1879 by a chemist who didn't wash his hands properly after working with coal-tar derivatives, and who then noticed a sweet taste while eating dinner. Cyclamates were discovered in 1937 by a chemistry graduate student at the University of Illinois, Michael Sveda, who lit a cigarette without washing his hands while working with chemicals—again, an unexpected sweet taste alerted him to his discovery. Aspartame followed in 1965 when James Schlatter, a researcher at G. D. Searle, spilled some of the material he was heating in a flask. "At a slightly later stage, when licking my finger to pick up a piece of paper, I noticed a very strong, sweet taste," he later recounted.

By the 1970s, researchers were supposed to be a little more careful about hand washing and tasting unknown substances. But Shashikant Phadnis, a graduate student from India working in London, misunderstood when his supervisor told him to test a new chemical he had just synthesized—he thought he was supposed

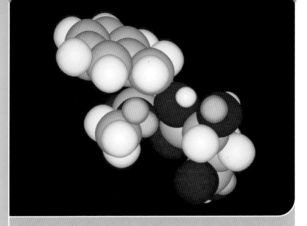

One of several sweeteners discovered accidentally, aspartame is composed of two amino acids, phenylalanine and aspartic acid, with a molecular structure as shown in this diagram.

to taste it. He did—and discovered sucralose, which is 600 times sweeter than sugar. (Aspartame is about 180 times sweeter, saccharin is 300 times sweeter, and cyclamates are 30 times sweeter.)

Since their introduction, artificial sweeteners have remained highly controversial because of concerns about their safety—but that hasn't stopped demand from continuing to expand. Each year around the world, about 100 billion pounds of artificial sweeteners are now consumed, creating a market valued at about $1.5 billion. And the latest entry to the market, neotame (approved by the Food and Drug Administration in 2002), was no accident: the compound, which is about 8,000 times sweeter than sugar, emerged from a concerted 20-year effort to find a new sweetener—deliberately, this time.

IN VITRO FERTILIZATION

During the final weeks of Leslie Brown's pregnancy, reporters besieged the hospital in the small town of Oldham, in northern England. One even phoned in a bomb threat in an attempt to gain access to her. Finally, on July 25, 1978, Mrs. Brown proceeded down a corridor lined with police and security officers to the operating room, where she delivered her baby by cesarean section.

The baby girl, Louise Joy Brown, was the first "test-tube baby," conceived through in vitro fertilization.

The birth was the product of a 12-year-long collaboration between physiologist Robert Edwards and gynecologist Patrick Steptoe, who in 1968 had become the first to successfully fertilize human eggs outside the body. Throughout the next decade, Edwards and Steptoe worked to produce a successful pregnancy, in the face of widespread criticism and fear about the ethical implications of tampering with the human reproductive system. Their first attempt at creating a viable pregnancy came in 1970, and its failure was followed by many others.

Finally, in November 1977, Steptoe removed an egg from Mrs. Brown, and Edwards fertilized it with her husband's

> "She took a deep breath. Then she yelled and yelled and yelled."
>
> PATRICK STEPTOE, on the moments after the birth of the first "test-tube baby"

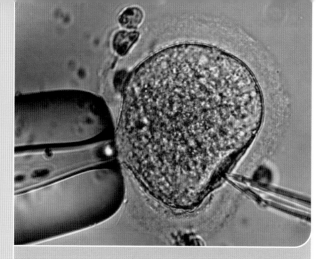

In vitro fertilization calls for the egg to be fertilized outside the body, then reimplanted in the mother's uterus. The first successful birth using the technique took place in 1978.

sperm. Two days later, the egg had multiplied into eight cells, and Steptoe reimplanted it in Mrs. Brown's uterus. The successful birth the following year ignited a rush of interest: more than one million children have been born using this technique since then.

Louise Brown has grown up happy and healthy, and in 2004 she married Wesley Mullinder. "We'd love to have children of our own one day," she said shortly after her wedding. "That would be such a dream come true."

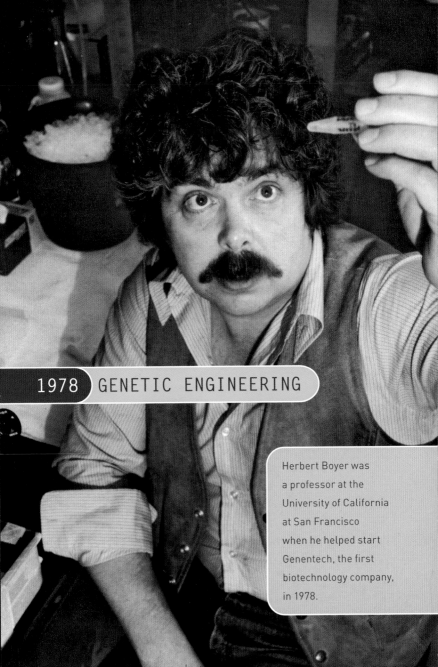

1978 | GENETIC ENGINEERING

Herbert Boyer was a professor at the University of California at San Francisco when he helped start Genentech, the first biotechnology company, in 1978.

THE AGE OF BIOTECHNOLOGY began in September 1978, when a small San Francisco start-up company called Genentech announced that it had implanted a synthetic gene into *E. coli* bacteria, causing the bacteria to produce human insulin. The company had been formed two years earlier, when a young venture capitalist named Robert Swanson had approached University of California at San Francisco biologist Herbert Boyer about commercializing his research.

In 1973, Boyer and a colleague at Stanford, Stanley Cohen, had been the first to splice genes from the DNA of one organism into another organism, but most scientists in the field believed that commercial applications were still many years away. Even so, Boyer agreed to give the venture capitalist 10 minutes of his time on a Friday afternoon—and the 10 minutes soon stretched into three hours at a local pub, at the conclusion of which each man laid down $500 to start their company.

> "We were so naïve we never thought it couldn't be done."
>
> HERBERT BOYER

At the time, insulin for diabetics was extracted from the pancreas glands of slaughtered cows and pigs—it took 8,000 pounds of pancreas glands to produce a single pound of insulin—so the company's success in producing human insulin from a test tube sparked a flurry of biotech activity. By the time Genentech went public in 1980, interest was so high that its stock rose from $35 a share to $88 in the first hour of trading. The company's insulin was approved for use in 1982, making it the first genetically engineered drug, and launching an industry valued today at over $400 billion.

1984) DNA FINGERPRINTING

You are unique. No one in the world has the same DNA sequence as you (unless you have an identical twin), so a readout of your DNA provides a foolproof form of identification. There's just one problem: there are more than three billion units in the human DNA sequence, so reading the entire sequence isn't practical. In 1984, British molecular biologist Alec Jeffreys came up with the solution, devising a way to compare only the parts of the sequence that show the greatest variation among people. By comparing these "mini-satellite" sections of DNA, scientists could determine whether two DNA samples came from the same person, two family members, or two unrelated people.

Jeffreys realized immediately that the technique could be used in forensic investigations, paternity disputes, and even in resolving immigration cases. He didn't have to wait long to try it out: shortly after news of the discovery was published in 1985, he was contacted by an immigration lawyer whose clients were Ghanaian citizens living in London. Their son had been detained after a visit to Ghana and accused of being an imposter—but Jeffreys was able to use their DNA samples to show that the boy was in fact the couple's son.

"I took one look, thought 'what a complicated mess,' then suddenly realized we had patterns."

ALEC JEFFREYS
on the moment in September 1984 when he developed the X-ray showing the first DNA "fingerprints"

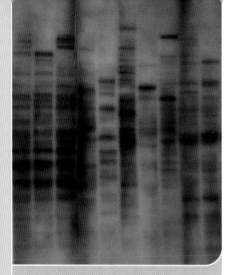

The first DNA fingerprint, prepared by Alec Jeffreys at Leicester University in the United Kingdom on September 19, 1984.

The following year, the technique was used in a criminal case for the first time, with surprising results. A young man who had confessed to a 1986 murder was also suspected of an earlier 1983 murder. Jeffreys used DNA to confirm that indeed, the same man had committed both crimes—but it wasn't the man who had confessed. The police then took blood samples from 5,000 men in the local community, and they used the resulting DNA evidence to identify the killer and release the man who had made the false confession.

Shortly after Kary Mullis conceived of the polymerase chain reaction (PCR, the technology that is the basis for the plot of *Jurassic Park* and a multitude of real-life DNA applications), he was presenting his work at a scientific meeting. He carefully explained the process to Nobel Prize–winning biologist Joshua Lederberg, who seemed amused. Mullis later recalled, "I think that Josh, after seeing the utter simplicity of PCR, was perhaps the first person to feel what is now an almost universal response to it among molecular biologists and other DNA workers: 'Why didn't I think of that?'"

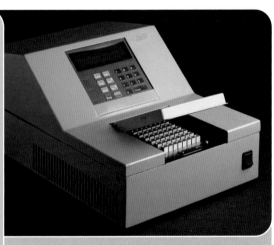

Polymerase chain reaction machines like this 1986 prototype offer a simple, quick, and inexpensive way to create billions of copies of a DNA fragment.

Mullis's insight may have seemed obvious in retrospect, but it was an unprecedented leap forward in DNA technology when it occurred to him, out of the blue, during a nighttime drive through the mountains of California. Using an enzyme called polymerase, PCR functions as an extremely powerful photocopier of genetic material. Starting with a single strand of DNA, you can generate 100 billion identical molecules in a single afternoon. "The reaction is easy to execute," Mullis points out. "It requires no more than a test tube, a few simple reagents, and a source of heat."

"Given the impact of the PCR on biological research and its conceptual simplicity, the fact that it lay unrecognized for more than 15 years after all the elements for its implementation were available strikes many observers as uncanny."

KARY MULLIS

The discovery was published in 1986, a year after Alec Jeffreys's invention of DNA fingerprinting (see page 86). PCR allowed forensic scientists to extract a minute amount of DNA from a drop of dried blood or a fragment of hair, then quickly create limitless copies of the genetic material in order to run multiple tests on it. Doctors, for example, could search for trace amounts of HIV genetic code to diagnose infection much sooner than by using conventional methods. And archaeologists could, in fact, retrieve and amplify ancient DNA, like the 395,000-year-old plant and animal samples found in Siberian ice in 2003. But—fortunately, perhaps—the ancient DNA does not survive intact, so the mixed-up fragments can't be used for Jurassic-style cloning.

1996 DOLLY THE SHEEP

The birth of a sheep named Dolly in a barn in Scotland in July 1996 did not attract public attention—because it was kept top-secret while a patent application was being processed. But Dolly became a media sensation seven months later, when it was finally revealed that she was the first mammal to have been cloned from the cells of another adult. The surprise result jolted biology into a new era, and it sparked widespread debate about the ethical implications of cloning animals—and, eventually, humans.

Ian Wilmut, who led the cloning team at the Roslin Institute in Edinburgh, had initially hoped to create genetically engineered farm animals that would produce therapeutic proteins in their milk, but his research led him instead to the cloning attempt. To create Dolly, DNA was extracted from the frozen udder cells of a long-dead sheep, injected into a hollowed-out egg taken from another sheep, and then zapped with electricity to fuse the two. Once the egg had divided several times, the cells were implanted in a surrogate sheep, which became pregnant.

Since then, more than a dozen species have been cloned, and a commercial pet-cloning service even operated from 2004 to 2006, delivering genetic copies of beloved

> "Dolly was a bonus. Sometimes when scientists work hard, they also get lucky, and that's what happened."
>
> IAN WILMUT

Dolly the Sheep gained instant fame and sparked ethical debates around the world when it was revealed that she had been cloned from an adult—a first for mammals.

pets for $50,000 each. But by the time Dolly died in 2003 (she was euthanized after contracting a painful lung infection), cloning techniques still had a very low success rate, with most attempts resulting in death or serious problems at birth.

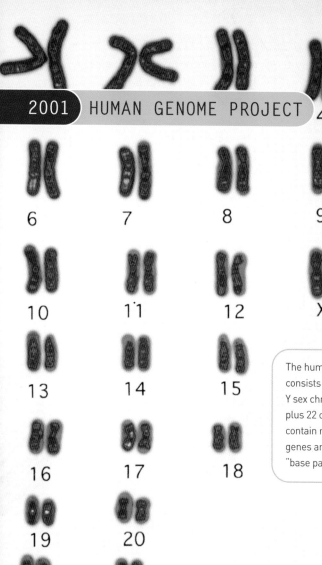

2001 HUMAN GENOME PROJECT

4

5

6

7

8

9

10

11

12

X

13

14

15

16

17

18

19

20

21

22

—

—

Y

The human genome consists of the X and Y sex chromosomes plus 22 others, which contain roughly 25,000 genes and three billion "base pair" units.

THE QUEST TO SEQUENCE the human genome—to decode, in order, the three billion entries contained in the "book of life"—took a dramatic turn in May 1998. At that time, controversial scientist Craig Venter announced that his company would complete the sequencing in three years, at a cost of $300 million. The federally funded Human Genome Project had been working on the same goal since 1990, with a projected completion date of 2005 and a price tag of $3 billion. Suddenly, the race was on.

There are 24 chromosomes in the human genome (the X and Y sex chromosomes plus 22 others), each of which is essentially a very long strand of DNA made up of a sequence of "base pairs." Humans have about three billion base pairs, which make up the roughly 25,000 genes that encode information about everything from hair color to congenital disorders and susceptibility to diseases. Having the completed sequence allows scientists to, for example, associate certain genes with particular conditions and develop specially targeted therapies, although the full potential of "genomics," as this field is known, is only beginning to be unlocked.

The government genome program had its origins in the 1980s, when the prospect of sequencing the entire genome first began to seem plausible. The program that was eventually launched in 1990 took a methodical approach, starting with general maps of each chromosome showing where the genes were located, and gradually proceeding to the laborious task of sequencing individual base pairs. Venter's approach, in contrast, relied on "shotgun sequencing," in which the chromosomes were

> "Speed matters—discovery can't wait."
>
> CRAIG VENTER, giving the motto of his company

chopped into millions of disorganized fragments. These fragments then are sequenced individually, and patched back into the correct order by a supercomputer.

The scientists involved in the government effort initially doubted whether Venter's approach would work at all, but they stepped up the pace of their own efforts in response. The race to complete the sequence became increasingly acrimonious, until President Bill Clinton finally asked his chief science advisor, Neal Lane, to "fix it" and end the squabbling. Lane brokered a truce that saw the rivals publish draft versions of the human genome in competing scientific journals on February 12, 2001, thus ending the race in a grudging tie.

CHAPTER 05

According to legend, the idea for the instant camera came to Edwin Land during a vacation in Santa Fe, New Mexico, in 1943. Out for a walk with his wife and three-year-old daughter, he was snapping pictures with a Rolleiflex camera, when his daughter asked why she couldn't see the results right away. "Stimulated by the dangerously invigorating plateau air, I thought, 'Why not?'" he recalled many years later. "'Why not design a picture that can be developed right away?'"

By that time, Land was already a successful inventor, having created thin plastic sheets that "polarized" light, reducing glare in sunglasses and allowing the creation of mass-produced 3-D movie viewers. (He eventually obtained 533 patents, at that time second only to Thomas Edison.) During World War II, Land's company, Polaroid, was concentrating on making gun sights and surveillance tools for the war effort, but after the war ended in 1945, he returned to his daughter's challenging question. By 1947, he had developed a complex procedure that allowed the exposure, developing, and printing to take place inside the camera within about 60 seconds, and he demonstrated the prototype at a scientific conference that year. The Model 95 Polaroid Land camera was released in 1948. Retailing for $89.75, it produced a

"You always start with a fantasy. Part of the fantasy technique is to visualize something as perfect. Then with the experiments you work back from the fantasy to reality, hacking away at the components."

EDWIN LAND

Edwin Land's "instant" photography hit the market in 1948, with the introduction of the Model 95 Polaroid Land camera.

3 ¾ by 4 ¾–inch sepia-toned print; within a year, sales exceeded $5 million as consumers embraced the startling new technology.

1958 CREDIT CARD

In its second full year of operation, the BankAmericard (later renamed VISA) lost $9 million—which turned out to be a blessing in disguise. Bank of America had launched the card in 1958 by mailing it out to all 60,000 of its customers in Fresno, California, and quickly expanded its customer base across the state in 1959. The bank's competitors were watching closely, and they noted the heavy

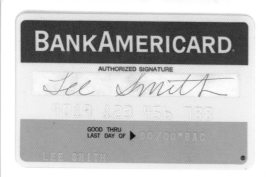

Launched in 1958, the BankAmericard was the forerunner of the VISA card. Its initial lack of success discouraged other banks from copying it, allowing the company to establish a dominant market position.

losses that accrued as merchants balked at the fees, and fraudulent or unpaid accounts turned out to be more frequent than expected. Not surprisingly, few serious competitors emerged.

The situation turned around in the mid-1960s. The number of participating retailers rose from 300 to 35,000, sales volume increased by 25 percent per year, and the bank reaped tens of millions of dollars of profit—a fact that was "one of the best kept secrets since the atom bomb," as Bank of America's head of consumer credit later recalled.

By the time other banks realized that the credit card was here to stay—and could be highly profitable—the BankAmericard had established a commanding lead of several years in the marketplace. Its success would soon be followed by the creation of MasterCharge in 1967, renamed MasterCard in 1979, the other leading credit card in the world.

Bank of America didn't invent the concept of the credit card. In fact, the term was coined in 1888 by Edward Bellamy in a novel describing life in the year 2000. In the early twentieth century, many stores issued cards offering credit to their own customers.

In 1949, Diners Club began offering a card accepted at a select group of New York restaurants, and it gradually expanded the concept to hotels and retail stores. The Diners Club was a charge card rather than a credit card: its balance had to be paid in full at the end of every month. Banks had also tried small local efforts, beginning with Flatbush National Bank's Charge-It scheme in 1947.

But it was Bank of America that succeeded in creating a credit card that was widely accepted, available to people who weren't customers of the bank, and offered "rotating credit" that required only a minimum payment each month. The result: more than 1.4 billion VISA cards have been issued, and they have processed more than $3 trillion in payments.

1960 PHOTOCOPIER

The story of the modern photocopier is extremely unusual. Unlike the vast majority of the inventions described in this book, it sprang fully formed from the mind of a single inventor. It relied on a completely new combination of scientific processes, rather than an incremental advance on previous technology, and nobody else in the world was working on similar designs. Had Chester Carlson, a young physicist working in a New York patent office during the

Although Chester Carlson demonstrated "xerography" in 1938, the first plain-paper photocopier, the Xerox 914, didn't ship until 1960.

Depression, not come up with his idea, it's entirely possible that no one would have developed the process that still powers most photocopiers and laser printers in the world today.

The "xerography" process relies on several fairly complex steps. A special surface in the machine is given an electric charge, making it sensitive to light. Light is reflected off the original document, transferring the pattern of bright and dark onto the charged surface. A fine powder (toner) is dusted onto the surface, collecting in the shape of the transferred pattern like iron filings in a magnetic field. Finally, the toner pattern is transferred to a new sheet of paper and heated to keep it in place. Each of these steps represented a significant technical challenge for Carlson, who conducted his initial experiments in his in-laws' kitchen and bathroom.

On October 22, 1938, Carlson and an assistant managed to produce the first xerographic image, which said simply, "10.-22.-38 Astoria." But it was another 22 years before the first automatic copier capable of printing on regular paper finally reached the public. Carlson approached more than two dozen major companies with his ideas but was met with what he later called "an enthusiastic lack of interest." Finally, a small Rochester company called Haloid gambled on the idea, eventually investing $12.5 million and more than a decade to develop it. Haloid's patience was rewarded: When the first Xerox 914 copier was finally shipped in 1960, it ushered in a new era—and made the company, which changed its name to Xerox in 1961, one of the biggest business successes of the 1960s.

> "As an invention, it was magnificent. The only problem was that, as a product, it wasn't any good."
>
> HAROLD CLARK, Xerox scientist, on the challenge of turning Chester Carlson's idea into a successful office product

1961 SOFT CONTACT LENSES

A purge of "politically unreliable" professors in Czechoslovakia cost Otto Wichterle his job at the Institute for Chemical Technology in 1958. With his unexpected leisure time, he was able to explore the uses of a new polymer he had developed called poly-HEMA hydrogel (HEMA stands for 2-hydroxy ethyl methacrylate). It was a transparent plastic that could absorb water without leaching dangerous molecules, making it an ideal material when required to be in close contact with human tissue. Over the next few years, Wichterle used a motor from an old phonograph and a children's Erector set to build a "spin-casting" machine to turn his new polymer into the first soft contact lenses.

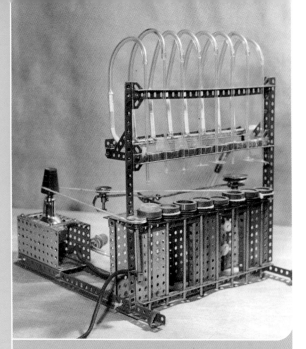

Otto Wichterle used an old phonograph motor and an Erector set to build the "spin-casting" machine that allowed him to turn a newly developed polymer in soft contact lenses.

Contact lenses had been around in one form or another since the nineteenth century, but they were made of rigid glass, Plexiglas, or plastic. In 1940, just 8,000 Americans used these "hard" contacts, which involved taking an impression of each eyeball to make a cast for the lenses, then hand-grinding them to produce the desired optical properties. The cumbersome lenses covered the whites of the eyes, and they cost more than $100 per pair.

California optician Kevin Tuohy improved the situation in 1948 with the invention of smaller, more comfortable lenses that rested only on the cornea, the central area of the eye.

Wichterle published his initial results in 1960, spurring interest from around the world. The patent rights to his new process were eventually secured by the U.S. company Bausch & Lomb, which introduced SofLens, the first soft contact lenses available to consumers, in 1971. The new lenses were more comfortable and let oxygen diffuse through the lens, reducing irritation and allowing the lenses to be worn for longer periods of time. By 2004, the market for contact lenses was approaching $4 billion a year, and soft-contact wearers outnumbered hard-contact wearers by more than two to one.

1969 QUARTZ WATCH

When the first digital watch was unveiled by Hamilton Watch Company in 1970, it was billed as the first new method of displaying time in 800 years. But a fancy display is no good unless the time it displays is accurate, so the 1969 release of Seiko's Quartz Astron 35SQ, the world's first quartz wristwatch, was arguably a more significant leap. It had an analog display, but its quartz mechanism kept time to within one second every five days— 100 times better than typical mechanical watches of the era.

Quartz is a crystal that oscillates thousands of times per second when an electric field is applied; researchers at Bell Laboratories had built the first quartz clock in the 1920s, keeping highly accurate time by counting the oscillations. But it wasn't until the 1960s that the miniaturization of electronics made it possible to squeeze all the components of a quartz clock into a wristwatch. Seiko's decade-long research program provided the official timekeeping at the 1964 Tokyo Olympics, and it produced a prototype watch in 1966.

Finally, in 1969, the Quartz Astron went on sale in Japan, and 100 were sold in the first week at an eye-popping price of $1,250. Inside the watch were a tiny quartz tuning fork oscillating 8,192 times per second; an integrated circuit (see page 18), whose 76 transistors and 29 capacitors had been hand-assembled and hand-soldered onto the circuit board; and a miniature stepping motor to turn the hands of the watch—the same basic mechanism still used in quartz watches today.

The first quartz wristwatch, Seiko's 1969 Quartz Astron 35SQ, attracted widespread interest, and its advent was even chronicled by the *New York Times*.

AUTOMATED TELLER MACHINE

After frittering away another lunch hour waiting in line to cash a check at his local bank, engineer Don Wetzel had an idea: why not build a machine to take over the simplest and most repetitive parts of a teller's job? He convinced his employers at Dallas-based Docutel to launch a $4 million program to build just such a machine. There were a series of technical challenges to overcome,

Chemical Bank's automated teller machine on Long Island was the first to offer 24-hour cash withdrawal, starting in 1969.

including how to automatically count bills with perfect accuracy, how to create a card that encoded customer information, and how to ensure security. The first machines had secure boxes with $5/8$-inch steel that would take eight hours to cut through with a blowtorch—overkill, Wetzel admitted later, but necessary to reassure the bankers, who "understood steel."

The remaining question was whether customers would be willing to trust the machines with their money. The first automated teller machine, or ATM, opened at a Chemical Bank branch on Long Island, following an advertising campaign that boasted, "On September 3, our bank will open at 9:00 and never close again!" The response was enthusiastic, and within less than a year, there were at least seven companies manufacturing ATMs.

Docutel's machine was not, however, the first to dispense cash. Earlier incarnations included a machine installed in 1967 at a Barclay's bank in England, designed by John Shepherd-Barron. The first version of the Barclay's machine dispensed cash in exchange for paper vouchers purchased in the bank; a second version accepted plastic cards but kept the cards and mailed them back later. Docutel's ATM laid the foundation for the modern ATM with a plastic card and personal identification number (PIN)—a format that has proved enduringly popular. By 2006, nearly 400,000 ATMs had been installed in the United States, processing over 10 billion transactions each year.

"Their mentality was: 'You have a check, I'm going to give you some money, and I hope you go away.'"

DON WETZEL on bank tellers, and why customers were willing to switch to ATMs

In the 1960s, Texas Instruments (TI) had a great technology—the integrated circuit (see page 18). However, its tiny market was limited to specialized military and industrial applications. The company's president, Pat Haggerty, thought the company needed a product that would bring the electronic revolution into every home (and boost sales as a result), so he asked Jack Kilby, the TI engineer who had coinvented the integrated circuit, to build a pocket-size electronic calculator.

Electronic calculators had been built as far back as the 1940s. IBM's Selective Sequence Electronic Calculator, for instance,

The CAL-TECH was a prototype pocket calculator built by Texas Instruments in 1966; a lighter consumer version was released in 1970.

occupied a room 60 feet long by 30 feet wide and contained 12,500 vacuum tubes. Better electronics brought somewhat smaller machines, which found a market with businesses and scientists, who until then had relied on slide rules for their calculations. The move from vacuum tubes to transistors (see page 14) finally allowed calculators to fit on a desktop; the first transistorized calculator, the Friden 130, appeared in 1963. Calculators still were not widely used, though: even the designer of the Friden 130, Robert Ragen, didn't use one. "From the transistor bias currents to the memory delay lines, I fleshed out all the circuitry on my Keuffel & Esser slide rule," he admitted.

With the additional miniaturization offered by the integrated circuit, Kilby and his team were able to build a prototype in 1966 that would fit into a large, sturdy pocket. The CAL-TECH, as it was dubbed, was made out of aluminum and weighed nearly three pounds. It was another four years before a consumer version of the calculator, the Pocketronic, hit the market, at a price of $400. It could add, subtract, multiply, and divide, and its case was now plastic, reducing the weight to less than two pounds.

Hewlett-Packard followed in 1972 with the first scientific calculator, the HP-35, which possessed more advanced math functions. To mark the occasion, it held a calculating contest pitting engineers with slide rules against engineers armed with HP-35s. The calculator team won, clocking in at 65 seconds against their competitors' five minutes, marking the end of one era and the beginning of another.

1979 SONY WALKMAN

In February 1979, Sony chairman Akio Morita called a meeting of engineers and planners at the company's headquarters in Tokyo. In his hand he held a compact tape player that had been specially modified to allow company founder Masaru Ibuka to listen to

The first Walkman was released in 1979, ushering in the era of portable music.

music during his long transatlantic business flights. The record function had been stripped away to save weight, and stereo sound capability had been added. Morita held up the modified device and said, "This is the product that will satisfy those young people who want to listen to music all day."

He was right. A production model of Sony's new Walkman was rushed to market less than five months later, and after a slow start, sales took off. Two years later, "Walkman" was beginning to appear in dictionaries as a synonym for portable stereo, and by 1995, sales had reached 150 million total units. Through its successors—portable CD and MP3 players—the once-novel concept of music to go is still keeping those young people satisfied.

There is a postscript: in 1980, a German-born Brazilian inventor named Andreas Pavel contacted Sony to ask for royalties. Pavel had developed a "stereobelt" in 1972, and though he didn't succeed in commercializing it, he had patented it in several countries starting in 1977. After two decades of litigation, Sony paid a multimillion-dollar settlement to Pavel in 2004, along with royalties from Walkman sales.

The old maxim that there's no such thing as bad publicity proved to be true for the company that released the world's second—not first—portable MP3 player, in 1998. The MP3 format, which allows digital music to be compressed into smaller files while maintaining

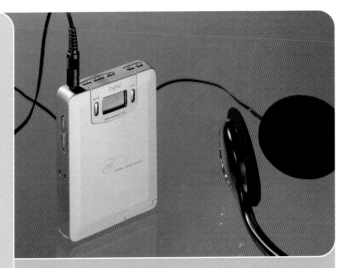

Saehan's MPMan F10/F20 was the first portable MP3 player on the market in the United States, followed a few months later by the Diamond Rio.

the quality of the sound, had been developed in the late 1980s at the Fraunhofer Institute in Germany. The first portable player to arrive on the market was the MPMan F10/F20 made by the Korean company Saehan, which had 32 megabytes of flash memory—enough to store about half an hour of CD-quality music. A few months later, the Diamond Rio became the second MP3 player released in the United States, retailing for under $200 and boasting the same amount of memory.

The advent of low-cost, portable digital music players struck fear into the hearts of the Recording Industry Association of America, which was afraid of piracy (with good reason, it turns out: CD sales peaked in 2000 and have been declining ever since). It decided to file a lawsuit to prevent the distribution of the new MP3 players, and since Diamond was headquartered in California, it was a much more convenient target than the Korean company, Saehan. Extensive media coverage of the legal proceedings gave Diamond a major boost, and by the time the case was settled in Diamond's favor in June 1999, Saehan was all but forgotten. In fact, many accounts now credit the Diamond Rio as the world's first MP3 player—three years before Apple's iPod changed the market forever.

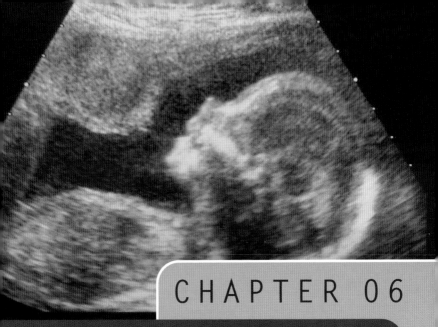

CHAPTER 06

POLIO VACCINE

In 1952, polio infected 58,000 Americans, killing 3,000 and leaving more than 21,000 paralyzed. It was the worst year ever for the disease, which is caused by a highly infectious virus that invades the nervous system and can cause total paralysis in a matter of hours. But Harvard's John Enders and his colleagues had sparked new hope a few years earlier when they discovered how to grow the polio virus in a test tube—a feat that later earned them a Nobel Prize—making it possible for several rival research groups to begin developing a vaccine. The race was on, but it proved to be a bitter one.

Researchers were split over whether a vaccine should use live or killed polio virus. Live- (but weakened) virus vaccines could be administered orally, and allowed immunity to be transferred from person to person, but carried a slight chance (about one in every 2.5 million doses) of infecting the recipient. Killed-virus vaccines required trained medical personnel to inject but were considered less likely to infect the recipient. Premature trials of both types of vaccine in the 1930s had infected a dozen children, killing six.

It was Jonas Salk, an immunologist at the University of Pittsburgh, who managed to create the first successful vaccine. His killed-virus vaccine was tested on 1.8 million children starting in 1954, and the results—80 to 90 percent effectiveness in preventing the disease—were greeted with great fanfare the following year.

"I've never studied a virus for its own sake. I've studied it for what relevance it had to human disease."

ALBERT SABIN, inventor of the oral polio vaccine

The first successful test of a polio vaccine, completed in 1955, paved the way for mass inoculations that have brought the complete global eradication of polio within reach.

University of Cincinnati researcher Albert Sabin, meanwhile, followed with a successful live-virus vaccine, which initially was greeted with indifference in the United States. He took it outside the country, and some 100 million children in the Soviet Union and its satellite countries were immunized with the Sabin vaccine in 1960. Although its success and ease of use eventually made it the dominant vaccine in the United States, the debate between live- and killed-virus proponents—and between Salk and Sabin personally—didn't die down for decades.

The global tally of polio cases dipped below 800 in 2003, before rebounding slightly. With both types of vaccines still in use, an effort to permanently eradicate polio is currently under way.

ULTRASOUND IMAGING

Today's ultrasound images of a fetus kicking inside its mother's womb have their origins in the military and industrial research that flourished during World War II. The Allies used sonar to detect German submarines by sending ultrasonic waves—sound waves with frequency above the limits of human hearing—through the water and detecting any waves that bounced back. Similarly, ultrasound flaw detectors proved useful for detecting cracks or weaknesses in metal, making them valuable tools in shipyards. When the war ended, scientists around the world began to ponder whether ultrasound could be used to peer inside the human body.

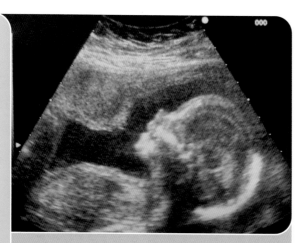

The ultrasound imaging systems that allow us to watch fetuses develop are descended from industrial tools used in the 1950s to find cracks in the hulls of ships.

One of the first scientists to explore ultrasound's diagnostic potential was George Ludwig, a doctor at the Naval Medical Research Institute in Bethesda, Maryland. He and his colleagues measured the reflected ultrasound signal from gallstones and other types of tissue, reporting their results in 1949. However, their experiments with living patients proved inconclusive. It was Ian Donald, an obstetrician in Glasgow, who made the next major advances. Through the husband of one of his patients, he gained access to an industrial ultrasound flaw detector, and in 1955 he used it to compare the signals from cysts removed from his patients with the ultrasound signal from a large piece of steak. By 1957, he was studying pregnant women, and had attained enough resolution to perform fetal head measurements, which eventually became a crucial marker of healthy development. He published his results in 1958, ushering in a new era of nonintrusive, nonradioactive imaging.

1957 FLEXIBLE ENDOSCOPE

Doctors have always wished they could see what's going on inside their patients, but the available technology hasn't always been up to the task. One of the earliest attempts to use an endoscope (a surgical instrument inserted into the body to provide an internal view) came in 1868, when a German internist named Adolf Kussmaul enlisted a professional sword swallower to choke down his primitive and rigid viewing device, an 18-inch-long brass tube. In the 1930s, a semiflexible endoscope was developed, but it remained awkward and difficult to use.

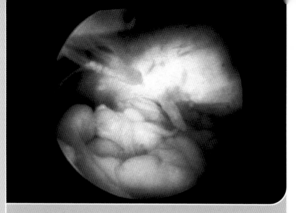

The ultrasound imaging systems that allow us to watch fetuses develop are descended from industrial tools used in the 1950s to find cracks in the hulls of ships.

In 1954, a physician at the University of Michigan named Basil Hirschowitz read a pair of journal articles about using bundles of tiny, flexible glass fibers to transmit images (see fiber optics, page 67). Intrigued, he flew to London to visit one of the labs where research was being carried out, and he returned to Michigan with rolls of the glass used to make optical fibers. Collaborating with members of the physics department, he spent the next few years trying to improve the transmission properties of the fibers. Finally, in 1957, a prototype of the fiber-optic gastroscope, inserted through the mouth to examine the stomach, was ready. To show that it could be administered without anesthetic or medication, Hirschowitz first tested the device on himself by swallowing it. By 1960, a commercial version was in production, and other applications for observing the body and performing minimally invasive surgery soon followed—much to the relief of patients who aren't sword swallowers.

IMPLANTABLE PACEMAKER

The pacemaker shown in this
chest X-ray applies a pulsed
voltage to stimulate a regular
heartbeat. About 200,000
pacemakers are implanted
each year in the United States.

IN 1951, ENGINEER WILSON GREATBATCH first heard about a condition called "heart block," in which electrical signals from the heart's upper chambers fail to reach the lower chambers, causing irregular heartbeats and sometimes death. His reaction: "I knew I could fix it." Greatbatch wasn't the only one to realize that electrical engineering could help correct the heart's poor signal transmission. In 1952, a Harvard surgeon tested the first external pacemaker, which applied a pulsed voltage to the patient's chest to stimulate a regular heartbeat. It worked, but the voltage caused skin burns and great pain to the patient. Later models sent the electrodes straight to the patient's heart but were still connected to a separate external unit, creating an infection risk.

> "I seriously doubt if anything I ever do will ever give me the elation I felt that day when my own two-cubic-inch piece of electronic design controlled a living heart."
>
> WILSON GREATBATCH, entry in lab diary, 1959

The solution was clear: an implantable pacemaker, powered by batteries, that could control the patient's heartbeat over the course of months or years while allowing the patient to leave the hospital. Greatbatch's first insight came when he grabbed the wrong resistor while wiring a device to monitor heart rate for animal psychology experiments at Cornell University. The result was a circuit that pulsed like a heartbeat, which Greatbatch incorporated into his pacemaker. He then miniaturized his device by using transistors—then a new technology—and enlisted the aid of Buffalo, New York, surgeon William Chardack. They successfully implanted the pacemaker in a dog in 1958, and in a human in 1960. By the end of the century, more than 200,000 lifesaving pacemakers were being implanted in U.S. patients each year.

1960 BIRTH CONTROL PILL

When the Food and Drug Administration approved a drug called Enovid for treating menstrual disorders in 1957, it added a warning: the mixture of synthetic progesterone and estrogen would also prevent ovulation. Two years later, more than half a million American women were taking Enovid—and not all of them had cramps. In 1960, the FDA approved Enovid for use as the first oral contraceptive.

The discovery that progesterone, a hormone secreted in the ovaries, could block ovulation had come 30 years earlier, but it

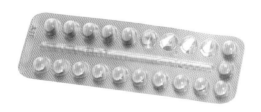

The first birth control pills were approved by the FDA in 1960, and they were soon used by millions of women.

took several major breakthroughs before a practical birth control pill could be developed. First, an eccentric chemist named Russell Marker quit his job at Penn State to set up a company in Mexico to produce progesterone from the wild yams he collected in the jungle, a discovery that lowered the price of progesterone from $200 to $3 per gram. Then in 1951, Carl Djerassi, often pegged as the father of the Pill, built on Marker's work by developing a synthetic equivalent of progesterone that was eight times stronger than naturally occurring progesterone and—crucially—could be taken orally. Still, years of research and massive trials, conducted initially in Puerto Rico and Haiti, remained.

When oral contraception was finally introduced in the United States, the reaction was predictably polarized: some viewed it as a sign of moral degradation, others as a way to stave off global overpopulation and put women in control of their own destinies. But the debate didn't slow the use of the pill. By 1964, four million women were taking Enovid, and there were six competing products on the market by 1965. And one effect was clear: in the decade following Enovid's introduction, the U.S. birth rate dropped by 30 percent, putting an end to the Baby Boom.

> "To be the very first to discover something gives you a very considerable amount of satisfaction. But if it were only that, clearly you should be satisfied without telling anyone about it. You're not. It would be the same thing as someone climbing Mount Everest for the first time and keeping his mouth shut about it."
>
> CARL DJERASSI

CORONARY BYPASS SURGERY

At a Cleveland hospital in May 1967, Argentine surgeon Rene Favaloro removed a long strip of vein from a patient's leg and grafted it to his coronary artery. This "bypass" operation, like a highway bypass, allowed blood to flow around an obstruction in the artery, stopping the patient's crushing chest pain. Coronary artery bypass grafting, as the procedure came to be known (often pronounced "cabbage" for its acronym, CABG), soon became a standard procedure for treating heart disease, helping to reduce heart disease deaths in the United States by 50 percent between 1970 and 2000.

Whether Favaloro was the first to perform the surgery has been a matter for debate. After he published his results, two American surgeons revealed that they had independently performed similar procedures as an emergency measure earlier in the decade, but they had neither reported their work nor attempted to develop the technique. Russian surgeons were also reporting early vein-grafting results in the mid-1960s. "So many people want to say they were first," Favaloro said later. "Let them. But the numbers of the Cleveland Clinic speak

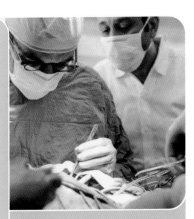

Surgical pioneer Rene Favaloro performed his first coronary artery bypass surgery in 1967 in Cleveland.

for themselves." Indeed, no one disputes the numbers: within a year of the first operation, Favaloro and his colleagues had performed 171 bypass operations and published their results widely, earning due credit as the pioneers of bypass surgery.

1971 CT SCANNER

Computed tomography (CT) is sometimes slyly referred to as the Beatles' greatest contribution to modern medicine. The first working prototypes of the machine (known initially as computed axial tomography, or CAT scanners) were developed at EMI Ltd.,

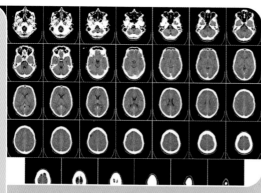

Computed tomography (CT) images like this scan of a human brain (from the base of the skull to the top) allow doctors to look at "slices" of tissue within the body, unlike the jumbled-together images produced by X-rays.

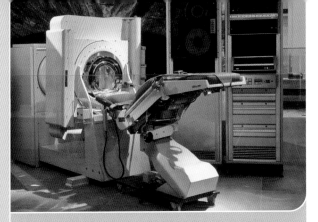

This CT scanner, designed by Godfrey Hounsfield at EMI, was used for the first patient trials in 1971.

a company flush with cash from the sales of Beatles records. Although EMI was primarily a music company, it also conducted pioneering electronics work from the 1940s until the 1970s, and the profits that accrued from the Beatles' success allowed it to invest heavily in long-term basic research.

Ordinary X-rays had first allowed doctors to peer inside their patients' bodies at the turn of the century, but the images they obtained were often muddied by successive layers of tissue. The CT scanner took X-rays from hundreds of different angles, then used a computer to process all the data and generate a much sharper image in three dimensions. The first prototype, developed by EMI's Godfrey Hounsfield, took 9 days to record images from 160 different angles, and then took another 2.5 hours of processing

"It looks exactly like the picture!"

The surgeon who operated on the first CT scan patient in 1971

time on a powerful mainframe computer to generate the image. The process was so slow that some of the initial specimens—animal parts from an abattoir—began to decompose while being imaged, producing gas bubbles that distorted the resulting pictures.

Hounsfield enlisted the aid of James Ambrose, a radiologist at a London hospital, to develop a machine for clinical use. In 1971, their prototype was used for the first time on a human patient, a woman suspected of having a brain tumor. The scan revealed a dark mass in her frontal lobe, which was confirmed afterward by the surgeon who operated on her. The technique was swiftly adopted, and by the time Hounsfield received the Nobel Prize in 1979 (along with Allan Cormack, a physics professor at Tufts University in Massachusetts who had independently developed a similar concept in the 1960s), more than 1,000 hospitals around the world had installed CT scanners.

1972 CYCLOSPORINE

Surgeons experimenting with organ transplants in the 1950s ran into an unexpected problem: even though they had the technical skill to transplant organs, the patients' immune systems would react to the new organ as a "foreign object" and reject it. An early kidney transplant succeeded in 1954—but it was between identical twins, so the recipient's body couldn't distinguish the new kidney from the old one. In the 1960s, scientists developed methods to temporarily suppress the immune system, and they began attempting liver and heart transplants. But even through the 1970s, the one-year survival rate for kidney transplants was just 50 percent.

Derived from a Norwegian fungus, cyclosporine (shown here with carbon atoms indicated in white, oxygen atoms in red, and nitrogen atoms in violet) suppresses the immune system's response to organ transplants.

The breakthrough came from a high, treeless plateau in Norway, where a fungus called *Tolypocladium inflatum* was found in the soil. The fungus was tested in 1972 as part of a routine screening program at the Swiss pharmaceutical company Sandoz, and a team of scientists led by Jean Borel and Hartmann Stähelin found that it had a rare combination: it could suppress immune system reactions without being toxic itself. Borel led the subsequent development of the drug, which was first used in kidney transplants in 1978—immediately increasing the one-year survival rate to 80 percent. Cyclosporine, as the drug became known, immediately revitalized the organ transplant field, which had nearly ground to a halt because of the immune system rejections, and remains in widespread use today.

Graduate student Lawrence Minkoff underwent to the first MRI scan of a live human in 1977.

ABOUT 60 MILLION PATIENTS around the world undergo magnetic resonance imaging (MRI) scans each year, enabling the diagnosis of everything from joint problems to brain tumors. MRI is one of the most significant medical inventions of the past century—but it's also one of the best examples of how hard it is to name a single "inventor" for a device that emerges from decades of work. When the 2003 Nobel Prize in medicine was awarded to Peter Lauterbur and Peter Mansfield "for their discoveries concerning 'magnetic resonance imaging,'" it wasn't the first Nobel related to MRI: the physics prizes in 1944 and 1952 also had been awarded for early work on magnetic resonance.

For decades, though, magnetic resonance was used mainly by chemists to study the composition of tiny samples. The turning point came in 1970, when a medical doctor named Raymond Damadian compared the magnetic resonance signal coming from healthy rat tissue with the signal from a sample of cancerous tumor cells. Monitoring how fast the hydrogen nuclei contained in the tissue's water molecules flipped over when a high-frequency magnetic field was applied, he was able to distinguish clearly between the healthy and cancerous tissues.

Over the next decade, several research groups raced to build a machine capable of scanning the organs of live humans. Again, it was Damadian who reached the milestone first, in 1977. He served as guinea pig for the first attempt, but the machine didn't seem to work. Eventually, a graduate student helping on the project suggested, "Maybe you're too fat," since the extra mass of tissue could absorb part of the signal. A skinnier assistant was enlisted, and the team spent five hours taking a rudimentary scan of his chest cavity.

Lauterbur and Mansfield, meanwhile, had been developing techniques independently to scan large areas much more quickly than Damadian's machine could. These ideas were eventually adopted, making MRI scans practical for medical applications. A further refinement of the technique by Richard Ernst earned yet another Nobel Prize in 1991, this time for chemistry. Damadian's omission from the Nobel citation enraged him, and in 2003 he undertook an aggressive campaign to have himself added alongside Lauterbur and Mansfield, taking out full-page ads in major newspapers across the country. It was a strange postscript to the story of an invention that, thanks to the discoveries of a long list of scientists, continues to play a vital role in health care and medical research.

"There is absolutely no reason to exclude Dr. Damadian and willfully perpetuate the shameless wrong that has been done to a man who is, because of his brilliant mind and indomitable spirit, one of the greatest living benefactors of patients worldwide and the last person who deserves to be needlessly injured by a prize committee or by anyone else."

From a full-page ad in the *New York Times* on October 20, 2003, paid for by the Friends of Raymond Damadian

When fluoxetine hydrochloride, the antidepressant better known as Prozac, was first introduced in the 1970s, the world didn't exactly sit up and take notice. The marketing department at drugmaker Eli Lilly told the scientists who had developed Prozac that psychiatrists were satisfied with the existing "tricyclic" depression medications, and that the potential market might be too small to justify the cost of clinical trials. When the team submitted their findings to the prestigious journal *Science*, their paper was rejected because it wasn't considered to be of sufficient general interest to merit publication.

In 1974, the research was finally published in another journal, *Life Sciences*, though it took until 1987 for Prozac to be approved

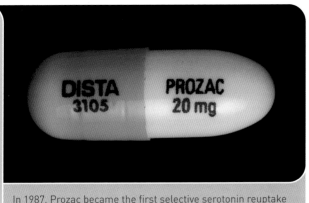

In 1987, Prozac became the first selective serotonin reuptake inhibitor (SSRI) approved for use as a depression medication in the United States.

for use in the United States. It was the first approval for a new class of drugs called selective serotonin reuptake inhibitors (SSRIs), which work by increasing the availability of the mood-altering chemical serotonin in the brain. These SSRIs, which include other drugs such as Zoloft, Paxil, and Celexa, have proved to be significantly more effective than previous antidepressants, with much milder side effects. As a result, Prozac quickly became one of the most prescribed drugs in the world, and since its approval, more than 54 million people have used it.

> "People have asked me how I can believe in God and also practice a science that affects mood and the human mind. I do not see this as a conflict. I believe that mental illness has a biological origin."
>
> DAVID WONG,
> coinventor of Prozac

Prozac has faced significant controversy over the years, including warnings that its use may be linked to suicidal impulses. Nonetheless, its widespread adoption has helped reshape our understanding of how personality and emotion can be chemically altered.

1995 PROTEASE INHIBITORS

Drug discovery in the modern world is an enormous and expensive endeavor, and the battle against AIDS (acquired immunodeficiency syndrome) in particular has enlisted huge teams of scientists from many different institutions. So it's fitting that the most successful weapons against AIDS so far, protease inhibitors, were developed simultaneously at several labs around the world.

When HIV (human immunodeficiency virus, which causes AIDS) infects a cell, it produces a long protein that is then chopped into shorter pieces as it exits the cell to spread the infection. The enzyme responsible for slicing the protein is called HIV protease; by the mid-1980s, researchers had realized that if they could find a way to interfere with the operation of this enzyme, they could slow down the infection. After a decade-long research effort, the Food and Drug Administration approved the first protease inhibitor, saquinavir mesylate (Roche's Invirase), in December 1995. Several other protease inhibitors developed by other

> "This is a problem for the world, and therefore we're going to solve it."
>
> DAVID HO

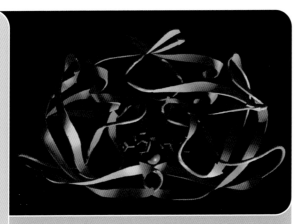

Protease inhibitors like the one whose structure is shown here by a computer-generated model play a key role in "cocktail therapy" for AIDS treatment.

companies were approved the following year, and researchers soon found that a "cocktail" combination of several drugs was most effective against the virus.

The effect was dramatic: after years of sharp increases, the number of HIV-related deaths in the United States dropped by 26 percent in 1996, and by a further 47 percent in 1997, thanks to the new drugs. Several studies showed that the virus could be reduced to undetectable levels in up to 90 percent of patients. David Ho, one of the pioneers of combination therapy (better known as HAART, or highly active antiretroviral therapy), was named *Time* magazine's Man of the Year in 1996. It wasn't a cure for AIDS, but for the first time, patients around the world could think about ways to live with the disease rather than simply preparing to die.

CHAPTER 07

>>> TRANSPORTATION

1953 AIR BAG

On March 12, 1990, two Chrysler LeBarons met in a head-on collision on Rural Route 640 near Culpeper, Virginia; both drivers walked away with minor injuries. It was the first time two cars in a collision had both deployed air bags to protect their drivers—and it came nearly four decades after the first air bag patents were filed.

In 1953, engineering technician John Hetrick received the first U.S. patent for a "safety cushion assembly for automotive vehicles." Hetrick's idea had been sparked by an accident in 1944 when he was repairing a Navy torpedo: a sudden release of the compressed air that powered the torpedo had instantly inflated the torpedo's canvas cover and shot it to the ceiling. Hetrick and German inventor Walter Linderer (who came up with similar ideas independently) each proposed air bags that would be inflated by compressed air—a design that later proved to be impractical. The problem: air bags needed to inflate within 30 milliseconds, and compressed air wasn't fast enough.

In the late 1960s, two key problems were solved. First, Allen Breed, a former RCA engineer who had started his own company, developed $5 electromechanical sensors that would tell the air bags when to inflate. The design was deceptively simple: essentially a steel ball in a tube held in place at one end by a magnet. The force of a collision would knock the steel ball free, allowing it to slide along the tube and make contact with an electrical circuit, deploying the air bag. Second, engineers began using sodium azide to inflate the air bag instead of compressed air: when the sodium azide was ignited, it released nitrogen gas quickly enough to fill the air bag within the 30-millisecond window.

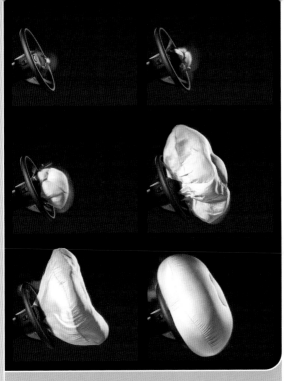

To be effective, air bags need to inflate within 30 milliseconds, a feat that is accomplished by igniting a chemical, sodium azide.

These advances led to enthusiastic predictions about air bag deployment, and the government proposed making air bags mandatory beginning in 1973—an idea vehemently opposed by automakers. As it turned out, neither the technology nor the marketplace was quite ready. General Motors offered air bags as an option on cars starting in 1974, but inconsistent performance and the added cost of several hundred dollars for the consumer meant that only about 10,000 units were sold, and General Motors withdrew the option in 1976. It wasn't until 1984 that Mercedes Benz began offering air bags, and by the end of the decade, the technology was firmly entrenched.

1956 CONTAINER SHIPPING

As a young trucker in 1937, Malcom McLean drove a load of cotton bales from Arkansas to New Jersey, then—as you had to in those days—sat around all day waiting for the dockworkers to finish transferring each individual crate from his truck onto a ship.

Standardized containers allowed shipments to be transferred directly between ships, trucks and railway cars without unpacking them, revolutionizing the shipping industry.

For large shipments, loading and unloading typically took more than a week at either end. It was a waste of time and money, and McLean wondered: why not simply transfer the entire trailer of his truck onto the ship with a crane?

It took nearly 20 years before McLean's idea finally came to fruition. By then, he'd built up his trucking business from a single rig to 1,700 trucks, but the logistics of loading and unloading the trucks at busy ports and railway terminals was still a major problem. McLean envisioned standard containers, the size of a truck trailer, that could be transferred directly between ships, trucks, and flatbed railway cars. But first he needed ships, trucks, and loading equipment specially fitted for the job—an extremely expensive proposition. Backed by hefty loans, McLean bought a steamship company in 1955 and began building.

In April 1956, his first container ship, the *Ideal X*, sailed from Newark with 58 aluminum containers. In Houston, the containers were unloaded and placed on 58 trucks that took the goods on to their final destinations—arriving faster than ever before. It was the beginning of what became known as the Container Revolution. In 1966, McLean opened a container port in the Netherlands and slashed four weeks off the best transatlantic delivery times. Before container shipping, transportation costs often accounted for more than half the consumer price for goods; after the new method gained acceptance, that contribution dropped to about 10 percent. Today, the largest container ships can carry eleven thousand 20-foot-long containers—equivalent to a 44-mile-long freight train.

> "It's often the people who know all about something that say it can't be done. I was totally ignorant, so I said, 'Why not give it a try?'"
>
> MALCOM MCLEAN

1957 ZEOLITE CATALYTIC CRACKING

When crude oil comes out of the ground, it contains a mix of hydrocarbon chains ranging from a few carbons atoms in length to hundreds of atoms long. Some lengths are more useful than others (the gasoline that runs our cars, for instance, contains chains that are 5 to 12 carbon atoms long), so crude oil is sent to refineries and processed to shorten or lengthen the hydrocarbons to the desired length. One of the key processes employed is "catalytic cracking," which breaks long hydrocarbon chains into shorter units using chemicals and heat.

In the 1950s, Charles Plank and Ed Rosinski were researching catalysts for Mobil Oil in New Jersey. Even though catalytic cracking had been in use since the 1930s, they had ample incentive to improve the process. "I well remember the statistic we used to quote in our early days," Plank later recalled. "If we could just increase the gasoline yield by 1 percent at the expense of gas and coke [two of the unwanted byproducts of cracking], it would be worth a million dollars a year to our company." Their goal was to improve the selectivity of the catalyst, so that the resulting hydrocarbon chains would be neither too short (gas) nor too long (coke).

The solution that Plank and Rosinski came up with was to use a material called zeolite, which is a clay-like mineral with

> "The tremendous interest of the whole petroleum industry in the zeolite cracking catalysts attests to their importance."
>
> CHARLES PLANK

Catalytic cracking units are a key part of oil refineries, upgrading heavy crude to lighter, more valuable products. The introduction of zeolite catalysts in the 1960s significantly increased the efficiency of the cracking process.

pores that are roughly the same size as the carbon chains to be cracked. This "perfect fit" means that only chains of the desired length will be cracked, dramatically increasing the selectivity of the catalyst. In 1957, the researchers showed for the first time that their approach would yield more gasoline from each barrel of crude oil, and by 1964, Mobil was using "Zeolyte Y" commercially. By the mid-1980s, about 35 percent of U.S. gasoline was produced by catalytic cracking, saving the country an estimated 200 million barrels of crude oil per year.

1958 JET AIRLINER

In addition to superior size, range and speed, the Boeing 707 introduced now-familiar features like folding armrests and opaque plastic window shades when it debuted in 1958.

AN EIGHT-HOUR, 41-MINUTE FLIGHT from New York to Paris in 1958 marked the debut of the world's first successful commercial jet airliner: the Boeing 707. The four-engine plane carried 181 passengers and cruised at 600 miles per hour, cutting the travel time in half compared with its closest propeller-driven competitors. Its range of 3,000 miles allowed it—just barely—to cross the Atlantic, with a stopover in Newfoundland.

The 707 became the prototype on which virtually all passenger jetliners have been based, with reconfigurable seats and thin wings swept back at a jaunty 35-degree angle. Its two key attributes were its speed and size, which reduced the cost of each seat per mile flown by a factor of two. Cheaper, shorter flights led to an explosion in passenger air travel during the following decade, transforming it into an accessible form of mass transportation. By 1965, an estimated 35 million people had flown along international routes.

Boeing wasn't alone in the race to develop a jet airliner. In fact, the de Havilland Comet had made the first commercial jet airliner flight back in 1952, but the craft was withdrawn two years later after three disastrous midair explosions attributed to metal fatigue. The 707, with its superior size and speed, managed to corner the market, and more than 1,000 were built before the design was retired in 1994.

In 1959, there was excitement in the air: "After a decade of research," *Business Week* reported, "it suddenly begins to look as if fuel cells might end up supplanting the internal combustion engine in automobiles." The reason for the enthusiasm was the demonstration of the first ever full-size fuel cell-powered vehicle—a tractor with 20 horsepower and 3,000 pounds of pulling power.

Fuel cells provide a highly efficient way to convert chemical energy into electricity; the most common designs use hydrogen as a fuel and oxygen as an oxidant, and they produce nothing but water and heat as exhaust by-products. The 1959 tractor, developed by a team at Allis-Chalmers Manufacturing in Milwaukee headed by Harry Ihrig, was just one of many fuel cell research projects being developed at the time, spurred by advances in fuel cell technology made by British engineer Francis T. Bacon beginning in the 1930s. In 1965, a General Electric fuel cell helped power the first manned Gemini mission in space, and the Apollo missions later in the decade used fuel cells made by Pratt and Whitney. But the technology remained too complex and expensive for consumer applications back on Earth.

"I believe that fuel cells will finally end the 100-year reign of the internal combustion engine. Fuel cells could be the predominant power source in 25 years."

WILLIAM C. FORD, JR., Ford Motor Co. Chairman, on the fuel cell, in 2001

This 1993 fuel cell–powered bus from Ballard Power is one in a long series of prototypes that have demonstrated the potential of fuel cell vehicles, without winning widespread adoption—so far. (Montreal's World Fair Expo '67)

Interest in fuel cell vehicles was rekindled in the early 1990s, led by Canadian company Ballard Power, which developed prototype fuel cell–powered buses that ran cleanly and silently—so silently, in fact, that when a glitch caused the first prototype to lose power during its unveiling ceremony in 1993, Ballard engineers were able to push the bus forward onto the stage without anyone noticing the problem. Later in the decade, demonstration fleets successfully served commuters in Vancouver and Chicago, but once again the price remained too steep for successful commercialization. For now, *Business Week*'s 1959 assessment of the fuel cell remains accurate: "It's portable, compact, and may even power autos someday."

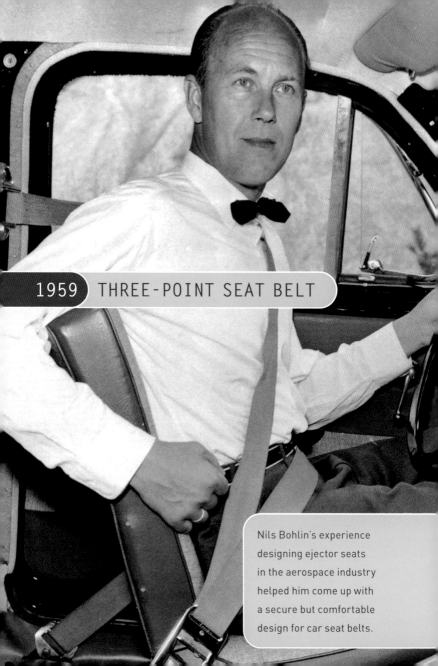

1959 THREE-POINT SEAT BELT

Nils Bohlin's experience designing ejector seats in the aerospace industry helped him come up with a secure but comfortable design for car seat belts.

NILS BOHLIN CAME TO VOLVO IN 1958 from the Swedish aircraft manufacturer Saab, where he had been designing ejector seats—so he knew all about strapping people in safely. But cars presented a different challenge: the existing lap belts were not only unpopular and uncomfortable, they also tended to exert an unhealthy, internal organ–crushing impact on their occupants during crashes. Newer V-shaped designs were being considered, but they were awkward and had anchor points connected to the seats, so that if the seat back flopped forward on impact (which often happened), the seat belt was useless.

Bohlin's new "three-point" design took less than a year to develop, featuring the now-familiar pair of belts—one across the chest and one across the hips—connected by a joint that could be buckled with one hand. Volvo made the belt standard in its cars in 1963, but acceptance was slower in the United States. In 1968, Bohlin came to the United States to make a presentation to the Consumer Products Safety Commission, and he showed a crash-test video with an egg strapped into the "crash cart." The demonstration of the new seat belt's light touch was convincing, and later that year, all new cars were required to include three-point seat belts. According to the Department of Transportation, seat belts now save more than 15,000 American lives each year, and Volvo estimates that Bohlin's design has saved more than a million lives worldwide since its release.

> "The pilots I worked with in the aerospace industry were willing to put on almost anything to keep them safe in case of a crash, but regular people in cars don't want to be uncomfortable even for a minute."
>
> NILS BOHLIN

1960 BYPASS TURBOFAN

The bypass turbofan is the improved jet engine design that made the first jumbo jet possible and now powers virtually all commercial jets. The basic principle of a jet engine, first patented by Frank Whittle in 1930, is simple: suck in air through the front of the engine, mix it with fuel, ignite it in a combustion chamber, and blow it out the back of the engine to produce forward thrust. If you want to push a bigger load or go faster, you need to draw in more air through the engine—but that takes more fuel, and you soon reach limits on how big a plane you can power and how far you can go.

The bypass turbofan draws in more air than a traditional jet engine, but some of the air bypasses the combustion chamber completely. The air is heated up on its way past the combustion chamber, then is expelled from the back, along with the ignited fuel, increasing thrust without gobbling fuel. Although the idea of a bypass turbofan had been patented by Whittle himself in 1936, it wasn't until 1960 that the technology finally made it into a jet airliner (see page 142)—a British Airways Boeing 707 powered by a Rolls-Royce Conway engine.

The Conway and other early bypass engines were not much more efficient than regular jet engines, since their "bypass ratio" (a measure of how much air bypassed the combustion chamber compared with how much went through it) was quite low.

"It is as old as nature— as old as, say, the squid. If you push air back, you push yourself forward."

FRANK WHITTLE, on the principle of the jet engine

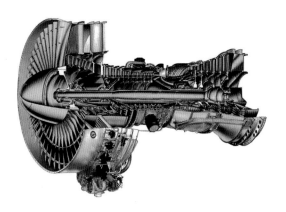

Pratt and Whitney's JT9D was the engine that made the first jumbo jet, the Boeing 747, possible in 1970, thanks to its high-bypass turbofan design.

But when Boeing began to consider the idea of a "jumbo jet" in the mid-1960s, it realized it would need engines with three times the thrust of the best engines of the time. The race to develop high-bypass turbofans ultimately led to the development of Pratt and Whitney's JT9D, which powered the new Boeing 747 when it started commercial service in 1970. The JT9D generated 42,000 pounds of thrust and sent five times more air through the bypass than through the combustion chamber. The result: since 1970, more than 3.5 billion people have traveled on a 747, and half a million more fly in one on any given day.

Unmanned aerial vehicles (UAVs) have a long history, but until the 1960s they were used mainly as flying bombs or for target practice. Then, in 1960, American pilot Francis Gary Powers was shot down while flying a spy mission over the Soviet Union, causing a major diplomatic incident—and kick-starting efforts to develop a long-range unmanned surveillance plane.

Four years later, a few weeks after the start of the Vietman War, a modified version of a target drone called the AQM-34 Ryan

Between 1964 and 1975, AQM-34 Ryan Firebees flew 3,400 unmanned surveillance missions over Vietnam. This photo shows a later version of the Firebee, being carried under the wing of a B-52 during a NASA test flight in 1977.

Firebee flew its first mission, traveling up the coast of southeastern China from Hainan Island to Taiwan. With a 13-foot wingspan, the diminutive Firebees could cruise above 60,000 feet altitude and at more than 600 miles per hour. They were launched from a C-130 Hercules mother ship, traveled up to 200 miles to take pictures of their targets, then flew back to friendly territory and deployed parachutes for retrieval. Between 1964 and 1975, more than 1,000 modified Firebees flew some 3,400 missions over Vietnam, with a success rate of 84 percent. The missions weren't just for surveillance—the drones were also used to fool antiaircraft systems into firing, revealing crucial information about enemy air defenses.

After the Vietnam War, UAVs fell out of favor, but they made a strong comeback during the 1990–1991 war in the Persian Gulf in spoofing the Iraqi air defense system, conducting surveillance and artillery spotting. Since then, they have become well established both for military applications and for civilian uses like environmental monitoring.

1975 CATALYTIC CONVERTER

When California issued the first car emissions standards in the early 1960s, engineers at Engelhard Corporation in New Jersey began considering the use of "catalytic converters" to turn pollutants like carbon monoxide and hydrocarbons into harmless water and carbon dioxide. They succeeded in building a prototype that ran successfully for an 80,000-mile test. But there was one problem: the devices didn't work if they were "poisoned" with lead, and virtually all gasoline at the time was leaded. The oil industry was one of the

Catalytic converters like this one, being tested at BASF Engine Laboratory in New Jersey, have eliminated billions of tons of pollutants worldwide.

major patrons of Engelhard's catalyst business, so the company decided not to rock the boat by publishing its results.

When Congress mandated new clean air rules in 1970, the playing field changed. The transition to unleaded gasoline was undertaken, and manufacturers rushed to develop a catalytic converter that would meet strict new standards by 1975. Ford became the first automaker to commit to the new technology, inking a deal with Engelhard in 1971. Three researchers at Corning Inc.—Rodney Bagley, Irwin Lachman, and Ronald Lewis—made a crucial breakthrough in 1973, developing an improved ceramic lattice for holding the catalysts that could withstand the high operating temperatures in a car engine without expanding and contracting excessively.

"People don't realize how bad it was when millions of cars were spewing out deadly fumes. Many of the cities were almost unbreathable."

RODNEY BAGLEY
of Corning, on life before the catalytic converter

The first catalytic converters went into 1975 model year cars and have been fitted to more than 500 million vehicles worldwide since then, eliminating more than 10 billion tons of pollutants. Modern cars now emit more than 99 percent less carbon monoxide, nitrogen oxides, and hydrocarbons than the unmonitored cars of the 1960s. And one of the unintended side effects of the catalytic converter has been just as important: forcing us to switch to gasoline without lead, which in the late 1970s was finally shown to have harmful effects on human health.

1977 COMPUTER-CONTROLLED IGNITION

When General Motors introduced the MISAR ignition system in 1977 Oldsmobile Toronados, it marked the first time a microprocessor (see page 26) was used in a production automobile. The MISAR (Microprocessed Sensing and Automatic Regulation) system controlled spark timing in the engine, subtly adjusting it to handle different driving and load conditions while keeping

The 1977 Oldsmobile Toronado featured the MISAR ignition system, marking the first time a microprocessor was incorporated in a mass-market car.

emissions to a minimum. The computer brought engine control to a whole new level: even with an eight-cylinder engine running at 3,000 rpm, a microprocessor can evaluate engine data so quickly that it spends most of the split-second time between engine cycles doing nothing.

The first electronic control system in automobiles was for headlights, in 1952. But it wasn't until the 1970s, with new federally mandated emission controls and concerns about fuel economy, that rapid development of electronic control systems in cars began in earnest. Chrysler was the first to introduce electronic ignition as a standard feature, in its 1972 models, which eliminated the constant wear and tear of the engine points. As ever more electronic features were added to cars, a General Motors engineer estimated in 1977 that a typical car contained 16 electronic subsystems, 460 yards of cable, 83 switches, 27 fuses, and 108 connectors.

Starting with MISAR, the move to computer-controlled, solid-state systems helped to reduce this tangle of wires, and electronic control has since spread to all parts of the modern car. Computers guide transmission shift points, antilock brakes, traction control systems, steering, air bag deployment, and a host of other functions. The result has been cleaner, more efficient, and safer automobiles.

1997 HYBRID-ELECTRIC CAR

When the first Toyota Prius rolled off the assembly line in 1997, it was about a century behind the cutting edge. Hybrid gasoline-electric cars had been common at the beginning of the twentieth century (Ferdinand Porsche's Lohner-Porsche, introduced at the

The billion-dollar investment Toyota made to develop the hybrid-electric Prius in the 1990s was seen as a gamble at the time, but it has proved to be a major commercial success.

Paris World Exposition in 1900, was one of the first), but battery power eventually lost the battle for market share to the internal combustion engine. By the 1990s, major car manufacturers were experimenting with all-electric vehicles and advanced fuel cell designs, but the hybrid approach was off the table.

It was the proposal of much stricter emissions requirements in California in the early 1990s that kick-started a Toyota program to design a more efficient car. When Toyota engineers ran into obstacles with their designs for an electric car, they started to consider a hybrid. In 1995, the company made the bold decision to

"We didn't think there was zero chance of success in meeting the goals outlined for us, but the feeling was that it was somewhere less than a 5 percent probability of succeeding."

TAKEHISA YAEGASHI, on the challenge of developing the Prius

proceed with a production version of the hybrid Prius—a gamble it backed with an investment estimated to be as much as $1 billion. The head of the Prius team, Takehisa Yaegashi, later recalled, "Feasibility studies and a lot of fundamental research were clearly needed, but instead we were told to jump right into mass production of a vehicle for consumers." When the team was told to have the car ready for production within two years, he said: "Is it any wonder we doubted?"

The engineers struggled to fit a large battery, an electric motor, and a gas engine into a small car, using computer models to find the best fit. To safely handle the large voltages between the electric motor and the battery, they adapted technology from Japan's famous bullet trains. The result was a car that relies on electric power at low speeds, where gas engines are least efficient, and recaptures energy to charge the batteries whenever the car brakes—giving a fuel efficiency that's twice some of its competitors.

The first Prius appeared in December 1997. Until 2000, the model was sold only in Japan, because engineers doubted whether it had the ability to handle the extreme heat or thin mountain air encountered in the United States. In 1999 the Honda Insight became the first hybrid on the market in the United States, and its early success paved the way for the North American debut of the Prius the following year, which was greeted by long backlogs of orders as customers hastened to embrace the technology. A decade after the car's release, Toyota had sold more than one million hybrid vehicles.

CHAPTER 08

The earliest paints, daubed onto cave walls thousands of years ago, were water-based. Even in Colonial times, whitewash (lime mixed with water) was more commonly used than oil-based paint. By the twentieth century, however, oil paint dominated, thanks to its superior finish and durability, even though it was difficult to clean up, required flammable solvents to thin, and took days to dry. But because oils of all kinds were in short supply during World War II, interest in alternative types of paint revived.

The solution came from prewar research into synthetic rubber, led by companies like Dow Chemical, which had developed a large manufacturing capacity for a synthetic latex called styrene-butadiene. When the demand for rubber dropped at the end of the war, Dow scientists sought to make use of their latex manufacturing capacity by developing a paint based on styrene-butadiene. Latex paint arrived in stores in 1948, in the form of Glidden's Spred Satin, which

Glidden's Spred Satin was the first latex paint, selling 100,000 gallons in 1948, its first year on the market.

sold 100,000 gallons that year and 3.5 million gallons just two years later.

By 1952, an article in the trade journal *Industrial and Engineering Chemistry* noted that "so-called latex paints" had surpassed oil-based paints in popularity, thanks to myriad advantages. These included "ease of application, good leveling, ease of touch-up, quick drying, rapid development of a considerable amount of resistance to washing, absence of dirt penetration, easy cleaning, a pleasing soft sheen, great durability, and the possibility of making dark colors, which presently are so much the vogue."

1952 NUMERICAL CONTROL MACHINING

In September 1952, MIT's Servo Lab unveiled a three-axis, digitally controlled milling machine, with instructions fed in via punched paper tape. It was the first demonstration of a "numerical control" machine (the name was suggested in an MIT contest with a $50 prize). The concept would ultimately revolutionize the manufacturing industry, but its origins were already the subject of controversy.

John Parsons, the vice president of a small manufacturing company in Traverse City, Michigan, is the man credited with the insight that machine tools could be controlled by computer. The shape of a part would be stored as a series of points, and punch cards would instruct the machine to move from point A to point B and so on. Parsons and his chief engineer, Frank Stulen, began to develop such a milling machine, and in 1949 they were awarded a contract from the Air Force to build one to make airplane wings.

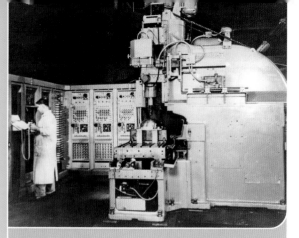

The first three-axis numerically controlled milling machine was unveiled in 1952 by the MIT Servo Lab.

Seeking a way of making his machine move automatically within an accuracy of 0.003 inches, Parsons subcontracted with the MIT Servo Lab to enhance that aspect of his machine.

The MIT engineers soon encountered practical problems—for example, to achieve the desired speed, punch cards would have to be read at the impossible rate of 500 per second. The MIT team modified Parsons's original concept in several ways, adjusting the computer program to simplify the calculations and replacing punch cards with paper tape. When the Air Force contract came up for renewal in 1951, MIT signed directly with the Air Force, eliminating Parsons Corporation from the project. Parsons was bitter: "I never dreamed that anybody as reputable as

"[Frank] Stulen did the calculation. I did the dreaming."

JOHN PARSONS

MIT would deliberately go ahead and take over my project," he later lamented. He did retain rights to the relevant patents, however, awarding a 12 percent share to the four principal MIT engineers.

By the time the prototype was unveiled in 1952, MIT had put its stamp on the project, making significant changes to improve its efficiency. Still, it was another decade before numerical control machining was widely adopted in industry. By then, the MIT engineers were taking the next logical step and adding computer-aided design (CAD) to the computer-aided manufacturing (CAM) that Parsons had first envisioned.

1958 CARBON FIBER

Carbon fibers are not a new creation—in fact, in the 1870s, Thomas Edison experimented with carbon fibers made from slivers of bamboo or threads of cotton as the filaments for his light bulbs. But the term "carbon fiber" means something different to us today: we think not of the fibers themselves, but of strong, lightweight composite materials made by combining carbon fibers with plastics.

The modern era of carbon fiber began in 1958 with a discovery by Union Carbide scientist Roger Bacon, who was exploring the high-temperature properties of the form of carbon known as graphite, as it switched between its solid, liquid, and gaseous states. Examining the carbon deposits that accumulated on one of his electrodes, he found a collection of "whiskers," each about one-tenth the diameter of a human hair. They were flexible, up to one inch long, and the strongest material (given their weight) that had ever been seen—more than 10 times stronger under tension than steel.

Strong, lightweight composites made from carbon fibers are used in applications ranging from golf clubs to space shuttles.

There was only one problem: "I estimated the cost of what it took to make them," he said later, "and it was $10 million per pound."

Over the next decade, scientists around the world found ways to produce the fibers more efficiently. One of the key advances came from Akio Shindo of the Government Industrial Research Institute in Osaka, Japan, who demonstrated in 1961 that high-quality fibers could be produced from a polymer called polyacrylonitrile, or PAN. Most carbon fibers produced today begin with PAN. Industrial production took off in the 1970s with the launch of carbon

"Will the first men rocketing to the moon wear graphite space suits?"

1959 press release from *Union Carbide*, after the discovery of high-strength carbon fibers. The answer was no, though every space shuttle has had carbon fiber parts.

fiber golf shafts and fishing rods, followed closely by aerospace applications. As prices have dropped, new uses continue to be found for this versatile material—from the crafting of lightweight car bodies to innovations in windmill turbines.

1959 FLOAT GLASS

British engineer Alastair Pilkington was helping his wife wash dishes one day in 1952 when the suds floating on the surface of the water caught his attention. Pilkington worked for the glass company Pilkington Brothers (though he was not, in fact, related to the company's founders), and the flash of insight that struck him that day revolutionized the glass industry—though not before nearly bankrupting Pilkington Brothers.

At that time there were two basic ways of making glass. Sheet glass was formed by taking a lump of molten glass and stretching it into a sheet, a quick and inexpensive process that was used to produce the vast majority of windows. But the stretching process produced slight differences in thickness throughout the glass, resulting in distortions—an effect you can still see in old buildings with original windowpanes. The alternative was plate glass, which was ground flat and polished using finer and finer sand. This laborious process made plate glass too expensive for anything but specialized applications like storefront windows.

Pilkington's insight was that the surface of a liquid is, by nature, perfectly flat. The "float glass" procedure that he developed involved floating a ribbon of molten glass in a bath of molten tin. As the molten glass floats down the bath of tin, it cools from 1,800 degrees

The inexpensive, distortion-free glass demonstrated by Alastair Pilkington (far left) revolutionized the glass industry following the introduction of his "float glass" method in 1959.

Fahrenheit to 1,100 degrees, at which point it is solid enough to be removed without marring the surface. The result: perfectly flat glass with a bright finish that doesn't need polishing—as good as plate glass, and as cheap to make as sheet glass.

Though Pilkington's initial tests in 1952 held promise, it took a full seven years of round-the-clock development before the first float glass production line started running, by which time the company had sunk £7 million into it—equivalent to a capital investment of nearly $1 billion today. Fortunately, it was a runaway success. Today float glass makes up about 90 percent of the global market for flat

"The principles of the process perhaps sound very simple, but there is a saying—I do not know who coined it—that the price of progress is trouble, and it is very true."

ALASTAIR PILKINGTON, on the long road between the invention of float glass and commercial success glass, with 320 production lines around the world churning out over one million acres of float glass a year—enough to cover nearly all of Delaware.

1961 CORDLESS TOOLS

It was changing demographics that led Black & Decker to develop the first cordless tools. Contractors who were installing aluminum storm windows (a booming market in the 1950s) were used to plugging their tools into an exterior light fixture, then getting the housewife inside to turn on the switch. But as more women entered the workforce, the contractors were sometimes left powerless when nobody was home.

Battery-powered tools were the obvious solution, but the technology for rechargeable nickel-cadmium batteries was still in its infancy. While professional plug-in drills at the time used more than 200 watts, Robert Riley's design team at Black & Decker could coax only 10 to 20 watts out of their battery packs. So they innovated, adjusting gear ratios to increase torque and reduce friction losses, and using highly conductive silver graphite to reduce electrical losses. The first cordless drill was ready for release in 1961, and Black & Decker unveiled it on NBC's *The Today Show*—with a dummy power cable that someone cut while host Dave Garroway was drilling into a piece of wood, prompting him to exclaim, "Is it magic?"

The cordless revolution put new power into the hands of the do-it-yourselfer—and soon after, into the gloves of astronauts. Observing Black & Decker's emerging expertise with cordless tools,

Black & Decker unveiled the first cordless drill in 1961, with a special appearance on *The Today Show*.

NASA asked the company to build a zero-torque wrench that could spin bolts at zero gravity without spinning astronauts, and a cordless drill to extract rock samples from the surface of the Moon.

1961 INDUSTRIAL ROBOT

A new worker joined the General Motors assembly line in Ternstedt, New Jersey, in 1961, taking over the dangerous and unpleasant job of picking up red-hot door handles and other parts that had just been die-cast from molten steel, then dropping them into cooling liquid. The new arrival was Unimate, the world's first industrial robot, and it was soon working around the clock, repeating the same task over and over with mind-numbing precision. Boasting a two-ton robotic arm and a grip with adjustable strength that could both lift eggs without breaking them and hoist 500-pound parts, Unimate could be programmed to repeat any sequence of movements its owners desired.

Unimate was the brainchild of George Devol, who had filed for a patent in 1954 for his idea of "universal automation, or 'Unimation.'" While Devol was the visionary, Joseph Engelberger was the practical businessman who helped turn Devol's visions into reality after the two men met at a cocktail party. Together, they formed a company called Unimation to sell the industrial robots. After the first Unimate proved itself in Ternstedt, GM followed up with an order for 36 more, but for the most part, American industry remained lukewarm to Devol and Engelberger's machine.

In Japan, on the other hand, Engelberger gave lectures to packed halls of fascinated businessmen. In 1968, Kawasaki Heavy

> "He had a patent for a programmable manipulator, and I said it sounded like it could be a robot."

JOSEPH ENGELBERGER, on his 1956 meeting with George Devol at a cocktail party

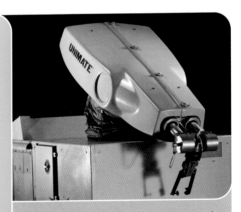

The first industrial robot, Unimate, started work in a GM plant in New Jersey in 1961. Shown is a later Unimate model, from 1979.

Industries licensed Unimation's technology, and soon became the world's leading industrial robot maker. Other Japanese companies like Fujitsu Fanac and Mitsubishi followed with designs based on Unimate, marking the beginning of a national infatuation with robots. Today, Japanese factories have three times as many industrial robots as American factories—thanks in part to their early interest in Devol and Engelberger's GM die-caster.

1965 KEVLAR

In October 2005, Atlanta police officer Corey Grogan was serving an arrest warrant when the suspect opened fire, hitting Grogan twice in the upper torso with a .45-caliber pistol. He survived, thanks to the body armor he was wearing, and in the process became the 3,000th member of the Kevlar Survivors' Club, a group of law-enforcement officers who owe their lives to an invention that had almost been discarded before its remarkable properties were discovered.

In 1965, DuPont chemist Stephanie Kwolek made an experimental polymer solution that turned out to be unexpectedly thin and cloudy and behaved strangely when stirred. Undeterred, Kwolek spun the solution into fibers and sent them for testing. The results showed that they were nine times stronger than anything she'd managed to produce before—an outcome

> "If things don't work out, I don't just throw them out, I struggle over them, to try and see if there's something there."
>
> STEPHANIE KWOLEK, on the persistence that led to her Kevlar breakthrough

Kevlar is five times stronger than the same weight of steel, making it ideal for lightweight body armor. This fabric is shown after impact by a fragment simulating a projectile.

so surprising that she sent the fibers back for testing several more times before revealing the results to anyone. The key: her new polymer produced straight, rigid fibers instead of the floppy, spaghetti-like strands that characterized the best man-made fibers of the time, like nylon and polyester.

The applications for the new fiber were not immediately obvious, despite the fact that it was five times stronger than the same weight of steel. DuPont released Kevlar in 1971, and ultimately it spent some 15 years and $500 million developing the technology. These days, Kevlar has hundreds of applications,

ranging from lightweight canoes to Mars-bound spacecraft—
though no one appreciates this material quite as much as the
members of the Kevlar Survivors' Club.

1968 WORKMATE WORKBENCH

Ron Hickman, a designer for sports car maker Lotus, was
sawing wood for a home renovation project in the early 1960s
when he realized that he had accidentally cut into the chair

This 1969 workbench made by Ron Hickman didn't spark
any interest from tool manufacturers, but Black & Decker
later licensed the design—and sold 60 million Workmates
in the next two decades.

he'd been using to support the wood. The incident spurred him to design a convenient workbench for his own use, with a surface made of two 5- by 2-inch lengths of beechwood that could be brought together as a giant vise. The frame consisted of lightweight aluminum and steel, and a footboard at the bottom allowed the unit to be stabilized by the user's foot. The whole bench folded up to the size of a suitcase and could be hung on a wall.

"The potential could be measured in dozens rather than hundreds."

1968 rejection letter from STANLEY TOOLS

After finishing the design in 1967, Hickman was convinced he had a product that other people would buy. But after making presentations to every manufacturer he could think of, he had nothing to show for his efforts but rejections. So he patented his design in 1968 and set up shop in an old brewery to manufacture them himself. By 1972, he was selling 14,000 units a year, mainly by mail order, leading Black & Decker to reconsider its earlier rejection. After licensing Hickman's design, it began mass production in Britain in 1973, and in the United States in 1975. The Workmate soon became a staple piece of equipment for do-it-yourselfers, and after 20 years of production, more than 60 million units had been sold.

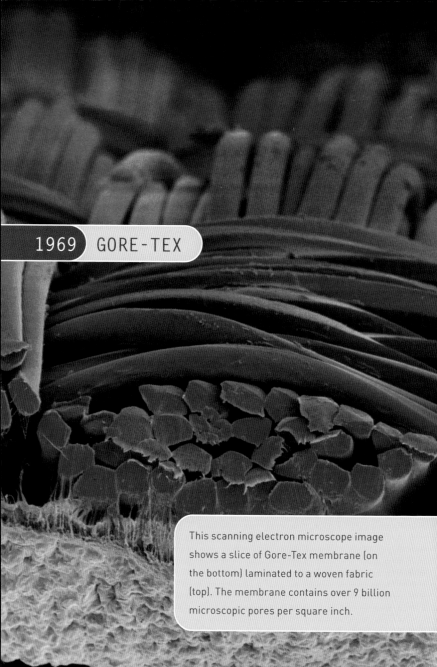

1969 GORE-TEX

This scanning electron microscope image shows a slice of Gore-Tex membrane (on the bottom) laminated to a woven fabric (top). The membrane contains over 9 billion microscopic pores per square inch.

A CAMPING TRIP IN WYOMING in 1970 was the site of Gore-Tex's first field test. Bill and Genevieve Gore hand-sewed a tent from their newly discovered fabric and took it on the road. On the first night of the trip, hail tore holes through the walls, and then rain filled up the tent like a bathtub. The accumulated water did have a bright side, though: it confirmed that the new material was waterproof, Bill Gore later recalled. "We just needed to make it stronger, so it would withstand hail."

The unique material at the heart of Gore-Tex had been discovered a year earlier. Bill Gore had been a chemist at DuPont in the 1950s, working on applications of the company's newly discovered Teflon (see nonstick pan, page 180), when he decided to form his own company, in 1958. Using Teflon to insulate electrical cables, Gore's business had grown so much that the equipment used during the first lunar landing in 1969 carried high-temperature cables manufactured by Gore. But he was still searching for new ways to use Teflon's unique properties—in particular, he wanted to figure out how to stretch and straighten the material's long, tangled fibers.

> "I figured that if we could ever unfold those molecules, get them to stretch out straight, we'd have a tremendous new kind of material."
>
> BILL GORE,
> on the quest to "stretch" Teflon that led to Gore-Tex

Gore and his son, Bob, spent hours heating rods of Teflon to various temperatures and gently trying to stretch them without breaking, but to no avail. Finally, one night in 1969, Bob was so frustrated that he yanked one of the rods angrily—and it didn't

Bob Gore demonstrates Teflon's ability to stretch when pulled quickly—the crucial step in the development of Gore-Tex.

break. Instead, it stretched into a form that was strong and highly porous, with tiny holes that permitted air to pass through but were too small for water molecules to penetrate. The new "expanded" Teflon was soon incorporated into Gore products like cables, filter bags, and, in 1975, vascular grafts for surgery. But it wasn't until 1976 that Seattle-based Early Winters became the first outdoor clothing manufacturer to include the waterproof-yet-breathable fabric in its tents and rainwear, launching Gore-Tex's most famous application. Over the next three decades, more than 100 million Gore-Tex products were sold.

CHAPTER 09

MICROWAVE OVEN

It was a candy bar melting in Percy Spencer's pocket that spurred the invention of the microwave oven. Spencer was a Raytheon Company scientist who had played a key role in World War II, developing techniques to mass-produce "magnetrons" to power the newly invented radar systems that gave Allied planes and ships a key strategic edge. Just after the war ended in 1945, he was in a Raytheon lab where a magnetron was being tested when he felt his candy bar soften. Other scientists had noticed the same thing happening, but it was Spencer who realized that the microwave radiation from the magnetron must be cooking his chocolate.

After some further tests—first popcorn, then an egg that exploded in the face of an unwary coworker—Spencer filed for a patent in October 1945. His original system called for food to pass through the microwave on a conveyor belt whose speed could be adjusted depending on how much cooking the dish needed. Other than that, his basic approach remains unchanged in today's models: microwaves generated by a magnetron cause water molecules in food to vibrate several billion times per second, heating them up. The warmth then spreads

"With the system described, I have found that an egg may be rendered hard-boiled with the expenditure of 2 kw.-sec [kilowatts per cm^3 per second]. This compares with an expenditure of 36 kw.-sec. to conventionally cook the same."

From PERCY SPENCER's patent application, submitted October 8, 1945

The first "Radarange" microwave ovens made by Raytheon in 1947 stood five and a half feet tall and weighed 750 pounds.

to the other parts of the food, while items containing no water, like dishes, aren't heated at all (at least not directly). As a result, microwave ovens are significantly more energy-efficient than gas or electric ovens.

Still, the first microwaves weren't a smash hit when they were released in 1947. The "Radarange" (the winning name in a Raytheon employee contest) was built like a fridge, standing five and a half feet tall and weighing 750 pounds—and cost more than $2,000, equivalent to more than $15,000 at today's prices. Though it was found to be useful in railway dining cars and ocean liners, it wasn't widely adopted. Even the first "home" microwave, released by Raytheon in partnership with the Tappan Stove Company in 1955, was hampered by a hefty $1,300 price tag. It wasn't until 1967 that Raytheon managed to shrink its new Amana Radarange to a more manageable size (15 inches tall, 91 pounds) and price ($495) and the age of convenience food began in earnest.

1951 SUPERGLUE

Harry Coover got two chances to discover a super-glue, and he didn't miss the second one. The first came when he was a young chemist assigned to find a material to make plastic gun sights during World War II. One of the materials he tried, an acrylic resin called cyanoacrylate, proved to be so sticky that it was almost impossible to work with—so Coover and his colleagues put it aside.

Nine years later, in 1951, Coover was the head of a group at Eastman Kodak assigned to develop a new material for

"Serendipity had knocked, but I didn't hear it."

HARRY COOVER on his first encounter with cyanoacrylate, in 1942

Eastman 910, as superglue was initially known when it was introduced in 1955, could hold at least 5,000 pounds—easily enough to hoist a convertible and its occupants, as it is doing here.

jet plane cockpits. One of his chemists, Fred Joyner, came to him and confessed that he had ruined a $700 prism, which had been glued stuck by the material he was testing: cyanoacrylate. This time Coover made the connection: cyanoacrylate might have potential as an adhesive. He immediately began testing it by gluing together whatever he could find. "Everything stuck to everything, almost instantly, and with bonds I could not break apart," he later recalled.

Cyanoacrylate adhesives first went on sale in 1955—the very first sale was to a government contractor assembling an atomic bomb, Coover said—and was known under various trade names: Flash Glue, Superglue, and Eastman 910. The glue's remarkable powers were introduced to the public when Coover went on the TV show *I've Got a Secret* in 1959 and used a single drop of glue to lift host Gary Moore off the floor. Since then, the range of uses has expanded dramatically. Famously, it was used by field medics in the Vietnam War as a form of spray-on instant bandage to halt uncontrolled bleeding, and variations on the original formula have since been approved by the Food and Drug Administration as a form of liquid bandage.

1954 NONSTICK PAN

Not surprisingly, the biggest challenge in making something out of Teflon is getting it to stick. That's why the first nonstick saucepans didn't appear until 1954—a full 16 years after DuPont chemist Roy Plunkett first discovered polytetrafluoroethylene (PTFE), the slippery substance that DuPont later trademarked as Teflon. Plunkett had stumbled across the new compound by accident while trying to develop an alternative to the chlorofluorocarbons used to cool refrigerators, but he quickly noted the unique properties of the substance. During World War II, PTFE was pressed into service by atomic bomb researchers to contain corrosive gases, and it was used in nose cones for conventional bombs. After the war, DuPont began searching for civilian applications.

In France, meanwhile, researchers had developed a way of attaching Teflon

> "Skillets were piled up, still in the shipping crates, as in a discount house, with the sales ladies handing them out to customers like hotcakes at an Army breakfast."
>
> ROGER HORCHOW of Neiman Marcus, describing the store's first sales of nonstick pans, in 1960

The biggest challenge in making nonstick pans was figuring out how to make the slippery Teflon stick to the metal surface of the pan.

to an aluminum surface: they roughened the aluminum with acid, covered it with Teflon powder, then heated it to create a smooth surface. One of the engineers, Marc Grégoire, was an avid fisherman, and he decided to use the process to coat his fishing gear to prevent tangles. When his wife heard about it, she suggested that he coat her cookware—so he did, with such good results that he patented the process in 1954. The Grégoires began selling nonstick pans from their kitchen; by the end of 1956, when they started the company Tefal, they were making 100 pans a day. Interest developed more slowly in the United States, with the first pans appearing only in December 1960, but they didn't take long to catch on: by mid-1961, stores were ordering a million pans per month.

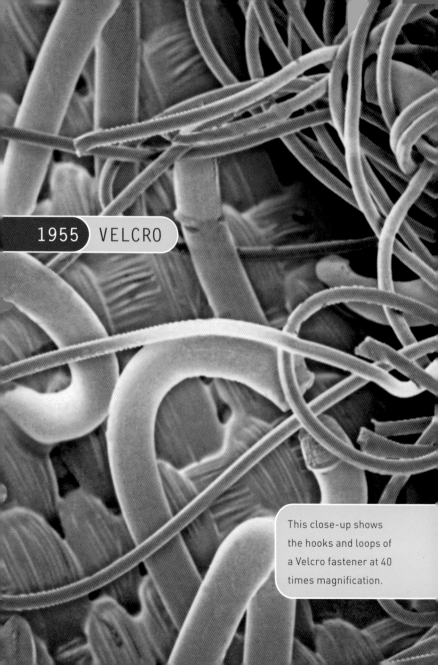

1955 VELCRO

This close-up shows the hooks and loops of a Velcro fastener at 40 times magnification.

THE INVENTION OF VELCRO was a classic example of "biomimetics"—taking inspiration from nature to produce something useful for humans. Georges de Mestral was a Swiss electrical engineer who enjoyed daily walks in the Alps with his dog. One day in the 1940s, he was struck by the annoying persistence of the burrs that adhered stubbornly to his pant leg and to his dog's coat. A longtime tinkerer and inventor (his first patent was for a toy airplane he designed when he was 12), he decided to examine the burrs under a microscope to understand how they stuck so firmly.

The secret of the burrs was the thousands of tiny hooks that terminated each spike, allowing them to latch onto the threads of his pants. That principle led de Mestral to develop the hook-and-loop fasteners we're familiar with today, although the initial attempts to sew thousands of tiny hooks and loops in a small area proved frustrating. He filed for a patent in 1951, and formed the Velcro company in 1952. The name came from two French words—*velours* (velvet) and *crochet* (hook)—and it's still trademarked, as the company fights to avoid letting Velcro become a generic term for hook-and-loop fasteners. These days, a two-inch-square piece might contain 3,000 hooks and loops, each a few hundredths of an inch high; with enough material, the fastener is sufficiently strong to support the weight of a 175-pound person hanging on a wall.

This handy device marks the official end of humanity's struggle for survival and the beginning of our quest for a really relaxing Saturday afternoon. The first wireless TV remote, designed in 1955 by Eugene Polley of what was then Zenith Radio Corporation, was essentially a flashlight. By pointing the beam of light at one of four light-sensitive spots on the TV screen, early couch potatoes

The Zenith Flashmatic, introduced in 1955, used beams of light to change TV channels and control volume. It was quickly supplanted by ultrasound- and then infrared-based remotes.

could adjust the volume and change the channels on their set. Unfortunately, direct sunlight could also change the channel, so Polley's "Flashmatic" had a short run.

Another Zenith engineer, Robert Adler, proposed the idea of using ultrasound for remote controls. The "Zenith Space Command," introduced in 1956, had four aluminum rods that were struck by tiny hammer mechanisms to generate sounds above the range of human hearing. Adding the necessary receivers to the TV set raised the price by about 30 percent, but consumers were more than willing to pay the cost of convenience, and other manufacturers soon followed suit. Ultrasound-based remotes persisted into the 1980s, to the chagrin of many a family dog. Modern remotes, which accompany 99 percent of new TVs and a wide range of other entertainment devices, rely on infrared frequencies to provide a robust—and quiet—channel surfing experience.

"People ask me all the time, 'Don't you feel guilty for it?' And I say that's ridiculous. It seems reasonable and rational to control the TV from where you normally sit and watch television."

ROBERT ADLER
in 2004, on his responsibility for creating a nation of couch potatoes

1961 DISPOSABLE DIAPER

off with
diapers

on with
PAMPERS!

Pampers—the discovery that makes diapers old-fashioned!

Here's why they keep your baby dry and happy as no diaper can!

Pampers' special stay-dry lining lets moisture through without getting soaked itself. The lining stays drier, helps keep wetness away from baby.

Moisture is trapped in absorbent layers below. Seven layers of softness make Pampers twice as absorbent as cloth diapers, ounce for ounce.

Flushabl proof out the rest a for baby washing

Introduced in 1961, Procter & Gamble's Pampers were the first mass-market disposable diaper.

New from Procter & Gamble! At food, drug, department and varie

SIZE MATTERS FOR DIAPERS. Like microchips, diapers have gotten steadily smaller—while achieving better and better performance—over the past several decades. Thanks to superabsorbent polymers that can soak up 300 times their weight in water, a modern disposable diaper can handle four "insults" without leaking or causing diaper rash, and it is much less bulky than its predecessors. But the history of the disposable diaper is surprisingly short: it was only in 1951 that a Connecticut housewife received a patent for a disposable diaper cover that was waterproof but breathable. Marion Donovan had cut her prototypes from shower curtains, and eventually she settled on parachute silk for the material—an idea that netted her $1 million when she sold the rights.

While the diaper cover kept sheets dry after a cloth diaper took an "insult," Donovan was unable to interest any major companies in her next idea, which was a fully disposable diaper filled with absorbent paper. Other researchers also experimented with the idea, but it wasn't until a decade later that Procter & Gamble's Victor Mills began testing prototypes on his grandchildren during long car rides. Mills was a prolific inventor, who during his tenure at P&G helped make Duncan Hines cake mixes less lumpy, Ivory soap bars more economical, and Pringles potato chips uniformly shaped. But it was Pampers, which in 1961 became the first mass-market disposable diaper, that remains Mills's most enduring legacy.

The next breakthrough in diaper technology came in 1966, when Carlyle Harmon of Johnson & Johnson and Billy Gene Harper of Dow Chemical filed independently for patents on the idea of diapers filled with small polymer flakes that unfurl when dunked in water. It took another two decades for these "superabsorbent" polymers to become standard, but they now fill some 95 percent of diapers used in the United States—making them small, comfortable, and nearly insult-proof.

1963 PULL-TOP CAN

The early steel beer cans of the 1950s had solid lids that required a "church key" opener, purchased separately, to gain access to the contents—a situation that was highly unsatisfactory if you happened to forget the opener. That's what happened to Ermal Fraze, an engineer and owner of the Dayton Reliable Tool and Manufacturing Company in Ohio, at a family picnic in 1959. After being forced to puncture the tops of beer cans using the bumper of his car, he turned his mind to a solution and came up with the first practical pull-top can. By pulling on a ring that acted as a lever, a prescored tab could be pulled off the top of the can. Fraze's design was unveiled in 1963 by the Pittsburgh Brewing Company, and just two years later, 75 percent of cans across the country sported the "self-opening" top. By 1980, sales of machine tools to make easy-open cans were generating $500 million a year for Fraze's company.

Success created a problem, however: the discarded pull-tabs became a serious garbage hazard, cutting bare feet at the beach, being swallowed by animals (and

> "I personally did not invent the easy-open can end. People have been working on that since 1800. What I did was develop a method of attaching a tab on the can top."
>
> ERMAL FRAZE
> in 1963

Today's easy-open drink cans follow the "Stay-On-Tab" design introduced by Dan Cudzik of Reynolds Metals in 1975.

occasionally by people who had dropped the tab into their can), and generally making a mess.

The struggle to design an easy-open can that would spawn no discarded pieces produced numerous proposals, some of which were briefly popular (like the two-hole pushbutton design pioneered by Coors). The safety challenges presented by sharp metal edges, as well as the manufacturing challenge of scoring the can tops to less than half of their 0.0088-inch thickness, were considerable. The design that finally caught on—and is still in use for beer and soda cans today—came from Dan Kudzik of Reynolds Metals in Virginia in 1975.

1969 SMOKE DETECTOR

Few people realize that the home smoke detector, a simple device available for less than $10 at hardware stores, is actually a piece of nuclear technology. At the heart of ionization-type smoke detectors is a small piece of americium-241, a radioactive element created as a by-product in nuclear reactors that emits a steady stream of alpha particles. These alpha particles travel across a gap in the detector and generate a tiny ionic current; however, if microscopic particles of smoke enter the detector, they disrupt the flow of ions, changing the current and triggering an alarm. This type of detector was first developed by Swiss engineer Ernst Meili in the 1940s, though it took years to perfect.

The other common smoke-detector design relies on a beam of light instead of alpha particles: if smoke particles deflect some of the light onto a photoelectric cell, the alarm is triggered. Research has shown that ionization detectors are quickest at detecting rapidly flaming fires, while photoelectric detectors are best for slow, smoldering fires. Whichever design you choose, there's compelling evidence for their effectiveness: more than 95 percent of U.S. homes

Battery-powered smoke detectors like this 1977 BRK Electronics model were a big hit starting in the 1970s, introducing the now-familiar low-battery chirp.

now have at least one smoke alarm, and 65 percent of home fire deaths occur in homes with no working smoke alarm.

While the technology has been around for decades, the turning point occurred in the late 1960s, when several companies raced to develop a battery-powered detector that would be easy to use in single-family homes. The first patent for a battery-powered detector went to Randolph Smith and Kenneth House of Norwalk, California, who filed their photoelectric design in 1966. Around the same time, Duane Pearsall of Statitrol Corporation in Colorado stumbled on a simplified design for an ionization detector when a technician's cigarette triggered a response in another device they were testing; Statitrol introduced its battery-powered ionization detector in 1970.

One of the most successful of the first alarms was the Early Warning Fire Detector from BRK Electronics, an ionization detector that went on sale for $175 in 1969. Its unique two-battery design used one battery to power the detectors and a second battery to monitor the first one. As the primary battery lost power, the detector would begin an intermittent chirp, at first once a day, and eventually once every few minutes—a mildly annoying reminder to change batteries that many people are still familiar with today.

1974 POST-IT NOTE

Spencer Silver, a research scientist with 3M, came up with a strange new adhesive based on acrylate copolymer in 1968. When he examined the adhesive with an electron microscope, it was composed of tiny microspheres that swelled to twice their normal size when they were dissolved in a solution, and that formed a pebbly layer like the surface of a basketball. The result was a unique adhesive with relatively

"When Post-its are still used after I am gone, it will be as if a part of me will live on forever."

ART FRY

low sticking power, but which could be unstuck or re-stuck repeatedly.

There was just one problem: no one at 3M could think of a good use for the new adhesive. Among the ideas they tried was a bulletin board covered in the adhesive, which was soon abandoned because it lost its stickiness as dust and dirt accumulated. The solution

Post-it notes, introduced in 1977, are a classic example of an invention—a low-power, reusable adhesive—that no one could initially find a use for.

didn't come until 1974, when another 3M scientist named Art Fry was confronting a problem while singing in his church choir: every time he opened his hymnal, all his bookmarks would fall out. The solution—sticky but nonpermanent place markers—made him think of Silver's glue.

Fry and his colleagues quickly developed what we now know as the Post-it note, but the public did not immediately grasp the idea. The notes were released in 1977 in four major markets but generated very little interest. The people at 3M, who had been using the notes internally ever since Fry developed them, realized that customers would need to try them to discover how useful they were. Starting with a test in 1978 in Boise, Idaho, they blanketed the city with free samples—and sure enough, people who tried the brightly colored notes loved them. They were released nationally using the same strategy in 1980, and by the late 1990s, studies showed that the average office worker received 11 Post-it messages a day.

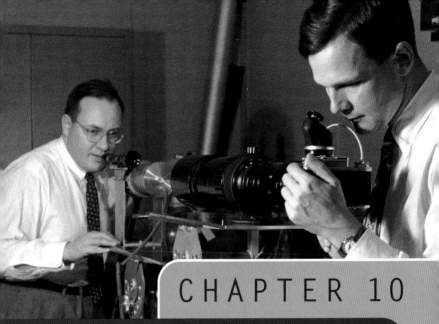

CHAPTER 10

The life-size holographic movies promised to us by science fiction writers have not yet materialized, but simple holograms like those used as security features on credit cards and banknotes are familiar to most people these days. The origins of the technology go back to 1947, when Hungarian-British physicist Dennis Gabor was trying to improve the resolution of electron microscopes. He devised an imaging technique that relied on the interference pattern between two beams of light to produce a three-dimensional image.

The leap from microscopic to full-size images had to wait for more than a decade, until the first working lasers (see page 204) were developed. Using the focused beam of light provided by a

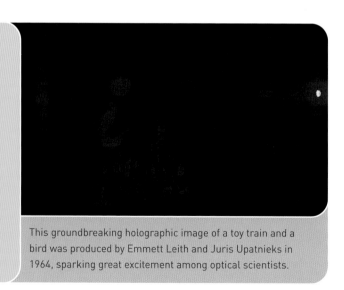

This groundbreaking holographic image of a toy train and a bird was produced by Emmett Leith and Juris Upatnieks in 1964, sparking great excitement among optical scientists.

laser allowed Emmett Leith and Juris Upatnieks at the University of Michigan to produce larger, three-dimensional holograms. They demonstrated their new technique by showing a hologram of a toy train at a 1964 meeting of the Optical Society of America, causing a sensation—optical scientists crowded around the demonstration, searching for hidden mirrors and wondering whether their eyes were being damaged by the laser. Similar research by Yuri Denisyuk in the Soviet Union soon allowed large holograms to be seen using ordinary light rather than lasers.

After the excitement of the early demonstrations in the 1960s, the pace of progress slowed. Artists such as Salvador Dalí began experimenting with holography, and industrial flaw detectors used the technique to check products such as airplane tires for imperfections. Another application proposed in the 1960s was holographic computer memory, using the three-dimensional properties of holograms for ultra-high-density data storage. Only now are commercial holographic memory systems appearing on the market, and they may end up being the most significant result of Gabor's insight.

1951 NUCLEAR POWER

Four 200-watt light bulbs glowing in a small building in the middle of a vast, uninhabited tract of Idaho desert marked the beginning of the nuclear power era. The scientists who developed the atomic bomb during World War II had realized that the same forces could be harnessed to produce electricity. To that end, a Nuclear Reactor Testing Station was set up on a former naval testing ground in

Idaho in 1949. Two years later, on December 20, Experimental Breeder Reactor-1 powered the light bulbs for the first time. The next day, the output was ramped up to 100 kilowatts, enough energy to power the whole building.

The success was a milestone, but not a surprise to the scientists. The research goal of the reactor was actually to develop a system that created more nuclear fuel than it burned—a so-called "breeder" reactor—since uranium supplies were very

When these four 200-watt light bulbs glowed on December 20, 1951, at Experimental Breeder Reactor-I in Idaho, it marked the first use of nuclear power to produce electricity.

limited at the time. That goal was achieved in 1953, although subsequent discoveries of large uranium deposits meant that breeder reactors never became widely popular. In 1954, a reactor in Obninsk, in what was then the Soviet Union, was the first to be connected to the local power grid and begin supplying homes and businesses with nuclear-generated power. Nuclear power remains controversial owing to safety and environmental concerns, but even so, there are now more than 400 commercial nuclear power stations operating in 30 countries.

"It was one of those things for which you have been preparing for months and years and you expect it to come through. When I turned the switch, I guess I was more interested in how the circuit breakers would function than I was in the significance of the test."

HAROLD LICHTENBERGER, project engineer for Experimental Breeder Reactor-1

1954 SOLAR CELL

In the early 1950s, AT&T was searching for a way to power parts of its telephone system in rural areas, where traditional batteries sometimes proved unreliable. They asked Daryl Chapin, a scientist at the company's Bell Laboratories research arm, to look into wind, thermoelectric, and steam power options—but Chapin suggested they also consider solar power. Researchers had been experimenting for decades with photovoltaic materials, which release electrons when struck by light, thus generating an electric current. But the best existing solar cells at the time, based on selenium, converted

The Bell Solar Battery was unveiled in 1954, with the initial goal of powering phone service in rural areas.

less than 0.5 percent of the energy they received from the Sun into usable power—an order of magnitude too low to be practical.

Meanwhile, two researchers in another Bell Labs research group had stumbled on a surprising property inherent in high-purity silicon. By introducing trace amounts of lithium and gallium into a piece of silicon, then shining light on it, Gerald Pearson and Calvin Fuller were able to produce five times more power than the output of the best selenium cells. After they shared their results with Chapin, the three scientists spent another two years overcoming practical difficulties in order to build a working device with 6 percent efficiency—the minimum threshold that would make solar power work for AT&T's phone system.

The new solar cell, dubbed the "Bell Solar Battery," was unveiled amid great enthusiasm at a press conference in 1954. A front-page story in the *New York Times* heralded "the harnessing of the almost limitless energy of the Sun for the uses of civilization." Bell's solar cells did make it into outer space aboard the first communications satellite, in 1962, but improving the efficiency and cost of solar cells has proven to be a long, slow process. Advanced solar cells have now achieved efficiencies of over 40 percent in the laboratory, but nearly 90 percent of the solar cell market is still based on the silicon technology pioneered by Chapin, Pearson, and Fuller.

"An old philosopher once said, 'It is stern work to thrust your hand into the Sun and pull out a spark of immortal flame to warm the hearts of men.' Yet in this modern age, men have at last harnessed the Sun, with the Bell Solar Battery."

BELL LABS short film announcing the first solar cells

1957 SPUTNIK

The man who gave the Earth its first artificial satellite was completely anonymous until his death—not only in the West, but also in his native Soviet Union. Sergey Korolev masterminded the Soviet program that launched Sputnik in 1957, convincing Nikita Khrushchev to let him pursue the project as an offshoot of the country's intercontinental ballistic missile (ICBM) program. But the space program and the identity of its top scientists remained

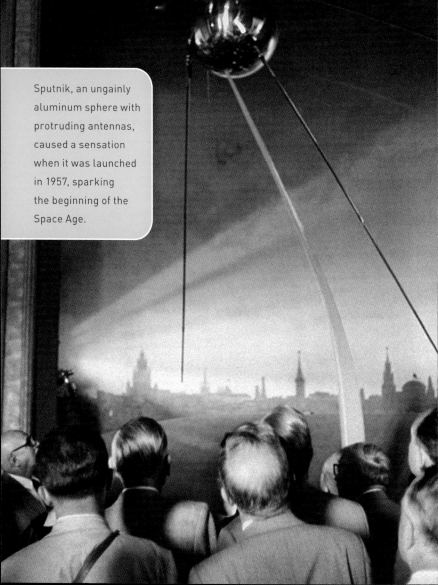

Sputnik, an ungainly aluminum sphere with protruding antennas, caused a sensation when it was launched in 1957, sparking the beginning of the Space Age.

state secrets: when Korolev died unexpectedly in 1966, the Soviet press waited until his identity was declassified three days later before finally reporting his many accomplishments.

Korolev's opportunity came in 1956, when the missile program hit a temporary snag. He took advantage of the hiatus and ordered his team to come up with a satellite design in just three months. The resulting plan was extremely simple: a sphere of aluminum alloy slightly larger than a basketball, with two radio transmitters and four antennas. Other researchers suggested adding some scientific instruments, but Korolev nixed the idea, fearing that the Americans would complete their satellite program first. The initial launch date was set for October 6, 1957—but then rumors surfaced that the Americans were planning a launch on October 5. The rumors were unfounded, but Korolev was taking no chances: he canceled some last-minute tests and moved the launch up to October 4.

The 184-pound satellite orbited the Earth more than 1,400 times. By the time it burned up in the atmosphere three months later, the Space Age had been launched, spawning a race to the Moon; satellites for communication, navigation, and weather forecasting; and a renewed focus around the world on science and technology.

> "The Earth is a sphere, and its first satellite also must have a spherical shape."
>
> SERGEY KOROLEV

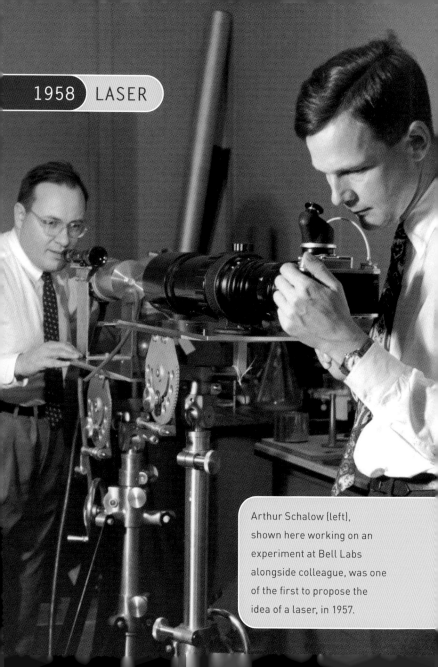

1958 LASER

Arthur Schalow (left), shown here working on an experiment at Bell Labs alongside colleague, was one of the first to propose the idea of a laser, in 1957.

NECESSITY ISN'T ALWAYS THE MOTHER of invention. When the first working laser was built in 1960 (laser is an acronym for light amplification by stimulated emission of radiation), it was viewed mainly as a curiosity that might eventually have some applications in communications and scientific research. "A laser is a solution seeking a problem," said the man who built the prototype, Hughes Aircraft Company scientist Theodore Maiman. That assessment quickly changed as scientists realized that the powerful, intensely focused beams of light produced by lasers were, in fact, the solution to a vast array of problems ranging from missile guidance to tattoo removal.

Several people made significant contributions to the development of the first laser—a fact that became significant when the tremendous commercial potential of the invention was realized. In the early 1950s, Columbia University professor Charles Townes invented a predecessor of the laser called the maser (microwave amplification by stimulated emission of radiation), which produced a beam of microwaves. In 1957, Townes and Bell Labs scientist Arthur Schawlow proposed replacing the microwaves with light, which would turn the maser into a laser. The pair began trying to build a prototype laser, and they received a patent for their idea in 1960.

> "I took it to a patent attorney, and he said, 'Well, what's the point in patenting if you don't see any uses for it?'"
>
> CHARLES TOWNES
> on the discovery
> of the laser

Meanwhile, also in 1957, a student at Columbia named Gordon Gould had independently developed an idea for the construction of a working laser (he was also the first to use the word "laser"). Gould mistakenly believed that he had to build a working prototype before he could file for a patent, so he quit his Ph.D. at Columbia and embarked on a quest to build one. He joined a company called Technical Research Group and obtained $1 million in funding from the military to produce a laser—but was then denied security clearance to work on the project himself, because he had taken part in a Communist study group in the 1940s.

In the end, Maiman beat both Townes and Gould in the race to build the first working laser. Maiman's device could operate only in short pulses, and it was extremely weak by modern standards (though it was a still a million times brighter than the Sun). But it was the original idea, not the first prototype, that was patentable. Though the scientific establishment (including the Nobel Prize committee) acknowledged Townes as the originator of the laser, Gould finally won a 30-year patent battle in 1988, granting him recognition for his role in the invention, along with millions of dollars in royalties.

1960 WEATHER SATELLITE

You might not want to bet your life on the accuracy of weather forecasts—but they're a lot better than they were before the first weather satellite, TIROS I (Television and Infra-Red Observation Satellite), was launched in 1960. The bird's-eye view of cloud patterns and storm movement that we now take for granted was unknown to meteorologists until the 3½-foot, 250-pound satellite was sent into orbit aboard a Thor-Able rocket in April 1960.

The key payload: two compact television cameras, two video

The first TIROS weather satellite being prepared for launch in 1960. The two TV cameras carried by the satellite can be seen in the immediate foreground.

recorders, and a rudimentary communications system that allowed it to transmit images back to Earth. Shortly after its launch, TIROS I imaged a previously undetected tropical cyclone off the coast of Australia, and during its 78-day mission it took nearly 23,000 photos of terrestrial weather patterns.

In the next five years, nine more experimental TIROS satellites were launched, as the National Aeronautics and Space Administration and the U.S. Weather Bureau sought to develop a global weather-observing satellite system. In 1965, TIROS IX was launched into an orbit that tracked the movement of the Sun, providing nonstop coverage of the entire sun-illuminated portion of the Earth. The following year, the first satellite in the new TIROS Operational System series was launched, paving the way for the current network of weather satellites.

Light-emitting diodes like this one have plunged in price from $260 in 1962, when they were invented, to just a few cents each today.

NICK HOLONYAK, JR., lost the race to build the first semiconductor laser by just a few weeks, beaten out by fellow General Electric scientist Robert Hall in 1962—but his consolation prize proved even more important. Unlike Hall's laser, which produced an invisible infrared beam, Holonyak's device produced red light that could be seen with the naked eye. By tweaking the parameters of his semiconductor alloy, he produced the first practical light-emitting diode, or LED—a durable, efficient light source that is now found everywhere, from the little red light that indicates coffee is brewing to the 22-story-high Reuters digital billboard in Times Square.

LEDs provided a simple way for computers and electronics to communicate with people, lighting up the display panel of the first personal calculators and digital watches. By the end of 1962, GE was offering Holonyak's LEDs for $260 each; the same gallium arsenide phosphide LEDs now sell for a few cents each. Researchers have since managed to produce LEDs of other colors, including white, which are now intense enough to function as car headlights. As the technology continues to improve, it looks increasingly likely that Holonyak was right when he predicted in a 1963 *Reader's Digest* article that LEDs would eventually replace incandescent bulbs.

> "I knew incandescent light was doomed a long time ago."
>
> NICK HOLONYAK in 2005, on the possibility that his LEDs would replace Thomas Edison's light bulbs

For the past few decades, superconductors have been heralded as an exotic technology that will lead to levitating trains and power lines with virtually no energy dissipation. Although those predictions have yet to come true, the remarkable properties of superconductors have been known for nearly a century. When cooled below a certain critical temperature, they permit electricity to flow with no resistance. And superconductors already play a crucial role in modern life, thanks to the widespread use of superconducting wire, which was first marketed by Westinghouse in 1964.

In 1957 a trio of Nobel Prize–winning physicists offered the first theoretical explanation of how superconductors work, sparking a surge of research in the area. By 1960, scientists at Westinghouse,

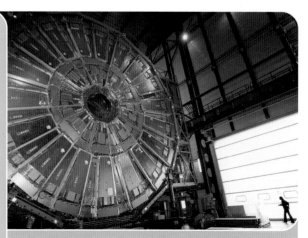

This enormous magnet at CERN, the European Organization for Nuclear Research, contains a coil of superconducting wire 28 miles long, making it the largest superconducting magnet in the world.

Atomics International, Bell Labs, and other research centers were experimenting with various superconducting alloys, seeking one that could be formed into a wire capable of carrying large currents. In 1962, the first prototypes of niobium-titanium wire were produced, leading to Westinghouse's commercial release two years later.

These days, superconducting wire—predominantly niobium-titanium—is most often found wrapped around large coils to produce powerful magnets; these magnets now power virtually all large MRI scanners (see page 128). The wire is also used for basic scientific research: the new Large Hadron Collider in Switzerland, the world's most powerful particle accelerator, also features the largest superconducting magnet, made by coiling a 28-mile length of superconducting wire weighing 16 tons. The discovery in 1986 of "high-temperature" superconductors has also led to new generations of superconducting wire, which manufacturers promise will soon enable the creation of an electric grid in which no power is lost due to electrical resistance during transmission.

1964 LIQUID CRYSTAL DISPLAY

In the fall of 1964, at an RCA lab, George Heilmeier applied a few volts to a thin film of butoxy benzoic acid sandwiched between two glass slides, then watched as the film changed from red to colorless. It was the first breakthrough using "liquid crystal," a peculiar form of matter existing somewhere between liquid and solid form. Heilmeier and his colleagues immediately grasped the discovery's potential. "The wall-sized flat-panel color TV was just around the

This 1968 RCA publicity shot shows the first liquid crystal digital clock—but it wasn't yet durable enough for consumer use, according to RCA's Robert Lohman, who appears in the picture: "The clock lasted just long enough to take a picture."

corner," he later recalled. "All that you had to do was ask us!"

The road to flat-panel TVs ended up taking much longer than Heilmeier, or anyone else, anticipated—but it started with his discovery of liquid crystal display (LCD) technology. The first prototype using the technology was a digital clock, built in 1967; RCA revealed its new technology to the world in a 1968 press conference. But by then it was clear that severe technological challenges would have to be overcome before flat-panel screens would be practical. RCA, as a dominant manufacturer of traditional cathode-ray tube displays, wasn't eager to develop a competing display technology, so it scaled

"The story of liquid crystal display technology has all the ingredients of a good novel. There was excitement, frustration, success, failure, and personal tragedy."

GEORGE HEILMEIER

back its research efforts. In Japan, on the other hand, the new LCD technology was greeted with enthusiasm. LCDs were small and extremely energy-efficient, making them ideal for portable applications. In the early 1970s, the first LCD watches appeared, and in 1973, Sharp introduced the first LCD pocket calculator.

Heilmeier's dream of a wall-size color TV was mostly forgotten for two decades, while scientists went on steadily improving LCD technology. By the late 1980s, full-color LCD displays for laptop computers were within reach. "Everything is beautiful except the price," quipped Takashi Shimada, head of Toshiba's Electron Tube and Device Division, in 1990. Fortunately, the price has continued to fall, and LCD displays are now a $100-billion-a-year market, dominated by the Japanese companies that were willing to invest in Heilmeier's technology back when a digital watch was still considered a pretty neat idea.

1969 CHARGE-COUPLED DEVICE

Like many great inventions, the charge-coupled device (CCD) has had a huge impact that its inventors did not originally anticipate. Willard Boyle and George Smith were researchers at AT&T's Bell Labs in New Jersey when they decided to have a brainstorming session in Boyle's office one day in 1969. They were trying to devise a new kind of computer memory when they sketched out a way of shuttling electrons around on a silicon chip. But once they'd worked out the idea, they quickly realized that it had potential as an imaging device. Within a year, Bell Labs researchers had used Boyle and Smith's CCD to build the world's first silicon-based video

Bell Labs scientists Willard Boyle (left) and George Smith (right) came up with the idea for the charge-coupled device in 1969.

camera. By 1975, they had developed a CCD camera with an image sharp enough to be used in broadcast television. These days, you'd be hard-pressed to find a camcorder—or a digital camera or telescope—without a CCD imager at its core.

CCD imagers rely on an effect first described by Albert Einstein in 1905: when light strikes a material like silicon, it knocks electrons loose. If you divide up a silicon surface into an array of pixels, you can record the image by counting how many electrons are generated at each pixel. Willard and Boyle's insight was to develop a quick and simple way of counting the electrons by shuttling them

"It was after maybe an hour's work. We went over to the blackboard and we had some sketching there. We went down to our models lab and made one."

WILLARD BOYLE, on inventing the CCD

toward a detector one pixel at a time. The result is an imager that can be sensitive to light sources as faint as one incoming photon per minute, making it ideal for telescopes that observe distant astronomical objects. And CCDs are a silicon-based technology, which means they have gotten steadily more compact and affordable over the years, making our current multimillion-pixel digital cameras possible.

1978 GLOBAL POSITIONING SYSTEM

The NAVSTAR (Navigation System with Timing and Ranging) Global Positioning System (GPS) was initially a military tool, with its origins in a 1960s program that used satellites to guide nuclear submarines. But when Korean Air Lines Flight 007 was shot down as it accidentally passed through Soviet airspace in 1983 (an incident that could have been avoided with better navigational equipment), President Ronald Reagan announced that the $5 billion system would be made available for civilian use when it was completed.

GPS relies on a constellation of 24 satellites (plus spares) that orbit the Earth twice daily. The first Navstar satellite was launched in 1978, but the full complement of 24 was not reached until 1994. Each satellite carries a set of atomic clocks that are accurate to within one second every 32,000 years, and it sends out a repeated microwave signal broadcasting its location and the exact time.

An observer on Earth with a GPS receiver can use the time and position data from a single satellite to calculate how far away that satellite is. With signals from three satellites, you can triangulate your own exact position; with a fourth satellite, you can determine your altitude.

Different branches of the military were developing competing satellite-based navigation systems in the 1960s and 1970s, and many scientists contributed to the system that emerged when the programs were united. It was Roger Easton, an engineer at the Naval Research Laboratory in Washington, D.C., who suggested and patented the idea of using synchronized clocks to pinpoint position. It was only in the year 2000 that President Bill Clinton granted nonmilitary users access to the highest-accuracy, unscrambled GPS signal. Now, more than 95 percent of GPS units sold are for civilian use, and inexpensive, handheld units can determine your location to within three yards.

The Global Positioning System relies on a network of at least 24 satellites orbiting the Earth twice daily.

The scanning tunneling microscope (STM) is basically a very sharp needle controlled by a very sensitive motor. When the needle, whose tip can be as narrow as a single atom wide, is brought close to a surface, electrons jump across the gap. By moving the needle across the surface of an object and monitoring how much electric current flows through it, scientists can construct a topographic map of the surface with single-atom resolution. The STM's development in 1981 earned a Nobel Prize five years later for IBM Zurich researchers Gerd Binnig and Heinrich Rohrer.

In the late 1970s, Binnig and Rohrer wanted to examine the atomic defects that exist at the surface of otherwise perfect crystals, in order to better understand the structure of solids. But when

IBM scientist Don Eigler showed that the scanning tunneling microscope could push individual atoms around by spelling I-B-M with 35 xenon atoms in 1989.

they began to look for equipment that would allow them to study these defects, they came up empty-handed. "There was nothing, so we had to invent something," Binnig recalled later.

The tool they developed became the first to enable researchers to probe features on the scale of single atoms—and it proved to be more than just a microscope. In 1989, another IBM scientist named Don Eigler discovered that he could use the sharp tip to manipulate individual atoms. To demonstrate this, he spelled I-B-M with 35 xenon atoms, creating one of the world's most famous scientific images and helping usher in the emerging era of nanotechnology.

"This process is just like Columbus going from Europe to America: on the way there, he has no clue that he is coming closer. We were in exactly the same situation because the instrument never worked."

GERD BINNIG, on the early experiments before the first successful STM prototype

SOURCES

Chapter 01: Computers

Transistor:

Harold Evans. *They Made America: From the Steam Engine to the Search Engine*, Little, Brown & Co., 2004.

Intel Corp., "Fact Sheet: Intel 45nm Transistors," 2007. Available at: http://www.intel.com/pressroom/kits/45nm/Intel45nm FunFacts _FINAL.pdf.

William Shockley. "Transistor Technology Evokes New Physics," Nobel Lecture, Dec. 11, 1956. Available at: http.//www.nobelprize.org/nobel_ prizes/physics/laureates/1956/shockley-lecture.pdf.

Transistorized. PBS documentary and interactive website. http://www.pbs .org/transistor/index.html.

Visible Storage exhibit. Computer History Museum, Mountain View, California.

Disk drive:

R. Bradshaw and C. Schroeder. "Fifty Years of IBM Innovation with Storage on Magnetic Tape." *IBM Journal of Research and Development 47*, no. 4 (July 2003).

Mark Ferelli. "The Most Important Invention of the 20th Century?" *Computer Technology Review* 24, no. 4 (April 2004).

IBM Corp. "Billions and Billions of Bytes: The Heritage of IBM Storage." IBM Archives. http://www-03.ibm.com/ibm/history/exhibits/storage/ storage_intro.html.

Integrated circuit:

Martin Campbell-Kelly and William Aspray. Computer: *A History of the Information Machine*. Basic Books, 1997.

Harold Evans. *They Made America: From the Steam Engine to the Search Engine*. Little, Brown & Co., 2004.

Intel Corp. "Intel's Transistor Technology Breakthrough Represents Biggest Challenge to Computer Chips in 40 Years." Press release, Jan. 27, 2007.

Gordon E. Moore. "Cramming More Components onto Integrated Circuits." *Electronics* 38, no. 8 (April 19, 1965).

Minicomputer:

Martin Campbell-Kelly and William Aspray. *Computer: A History of the Information Machine.* Basic Books, 1997.

"DEC PDP-8, Object 1989.0521.02," Computer History Collection, Smithsonian National Museum of American History, Washington, D.C.

Digital Equipment Corp. "Documenting DIGITAL," a history of DEC hosted at http://research.microsoft.com/~GBell, 1998.

Harold Evans. *They Made America: From the Steam Engine to the Search Engine.* Little, Brown & Co., 2004.

Interview with Dag Spicer, Senior Curator, Computer History Museum, Mountain View, California.

Computer mouse:

Christina Engelbart. "A Lifetime Pursuit." Biographical sketch hosted by the Bootstrap Institute, June 24, 2003.

Douglas Engelbart. "The Demo," a 90-minute video of live demonstration at the Fall Joint Computer Conference. Dec. 9, 1968, hosted at http://sloan.stanford.edu/MouseSite.

"Inventor of the Week: Douglas Engelbart." Lemelson-MIT Program (Jan. 2003).

ARPANET:

Martin Campbell-Kelly and William Aspray. *Computer: A History of the Information Machine.* Basic Books, 1997.

Internet History exhibit. Computer History Museum, Mountain View, California.

Barry M. Leiner, Vinton G. Cerf, David D. Clark, et al. *A Brief History of the Internet.* Internet Society, 2003. http://www.isoc.org/internet/history/brief.shtml.

Microprocessor:

Intel Corp. "How Microprocessors Work." Intel Education Initiative.

"Interview with Marcian (Ted) Hoff." In *Silicon Genesis: An Oral History of Semiconductor Technology*. Stanford University Libraries, March 3, 1995.

John Markoff. "For Texas Instruments, Some Bragging Rights." *New York Times*, June 20, 1996.

The Silicon Engine exhibit. Computer History Museum, Mountain View, California.

Personal computer:

Martin Campbell-Kelly and William Aspray. *Computer: A History of the Information Machine*. Basic Books, 1997.

Les Freed. *The History of Computers*. Ziff Davis, 1995.

Selling the Computer Revolution: Marketing Brochures in the Collection exhibit. Computer History Museum, Mountain View, California.

World Wide Web:

Tim Berners-Lee. World Wide Web Consortium. http://www.w3.org/People/Berners-Lee/.

Harold Evans. *They Made America: From the Steam Engine to the Search Engine*. Little, Brown & Co., 2004.

Barry M. Leiner, Vinton G. Cerf, David D. Clark, et al. *A Brief History of the Internet*. Internet Society, 2003. http://www.isoc.org/internet/history/brief.shtml.

PageRank:

John Battelle. *The Search: How Google and Its Rivals Rewrote the Rules of Business and Transformed Our Culture*. Portfolio, 2005.

Harold Evans. *They Made America: From the Steam Engine to the Search Engine*. Little, Brown & Co., 2004.

Google Inc. Corporate history. http://www.google.com/corporate/history.html.

Verne Kopytoff. "Google Surpasses Microsoft as World's Most-Visited Site." *San Francisco Chronicle*, April 25, 2007.

CHAPTER 02: LEISURE

Kernmantle rope:

Edelrid Inc. Corporate history. http://www.edelrid.de.

Extreme Textiles: Designing for High Performance exhibit. Cooper-Hewitt
    National Design Museum, New York, New York.

A. J. McLaren. "Design and Performance of Ropes for Climbing and Sailing."
    *Proceedings of the Institute of Mechanical Engineering. Part L: Journal of
    Materials: Design and Applications* 220, no. 1 (2006): 1–12.

Videotape recorder:

BASF Group. Corporate history. www.corporate.basf.com.

Peter Hammar, Curator, Ampex Museum of Magnetic Recording , Redwood
    City, California (now a part of the Silicon Valley Archives at Stanford
    University). Personal communication.

James E. O'Neal. "The Videotape Recorder Turns 50." April 12, 2006.
    Available at: http://tvtechnology.com/features/news/ 2006.04.12-n_the
    _video_tape_02.shtml.

Sony Group. "The Video Cassette Tape." *Sony History.* http://www.sony.net/
    Fun/SH.

R. M. Warner. "Earl Masterton: A Fresh Slant on Videorecording." *IEEE
    Spectrum*, Feb. 1996.

Stewart Wolpin. "The Race to Video." *Invention and Technology Magazine* 10,
    no. 2 (Fall 1994).

Polyurethane "foam" surfboard:

Anne Bell. "Surfers Hit Hard by Foam Company Closure." *PBS NewsHour
    Extra*, Dec. 27, 2005.

William Finnegan. "Blank Monday." *The New Yorker*, Aug. 21, 2006.

Hobie Inc. Corporate history. http://www.hobie.com.

Nat Young. *The History of Surfing.* Palm Beach Press, 1996.

Video games:

Stewart Brand. "Spacewar." *Vanity Fair*, Dec. 7, 1972.

Corie Lok. "The Start of Computer Games." Technology Review, June 2005.

David Pescovitz. "The Adventures of King Pong." June 12, 1999. http://www.salon.com.

"Spacewar!" *PDP-1 Restoration Project*, Computer History Museum, Mountain View, California.

Music synthesizer:

Frank Houston. "Robert Moog." April 25, 2000. http://www.salon.com.

Interview with Trevor Pinch, coauthor of *Analog Days: The Invention and Impact of the Moog Synthesizer* (Harvard University Press, 2004). Department of Science and Technology Studies, Cornell University, Ithaca, New York.

Allan Kozinn. "Robert Moog, Creator of Music Synthesizer, Dies at 71." *New York Times*, Aug. 23, 2005.

Karin Lipson. "Making Musical History." New York Times, June 17, 2007.

"Moog Archives," online site maintained by Roger Luther. http://moogarchives.com.

Digital music:

Adam Holdorf. "The Discoverer." *Reed Magazine*, Nov. 2000.

International Federation of the Phonographic Industry. "2000 Recording Industry World Sales." http://www.ifpi.org/content/ library/worldsales2000.pdf.

"Inventor of the Week: James T. Russell." Lemelson-MIT Program, Dec. 1999.

Philips Corp. "Optical Recording: Collaboration With Sony." Philips corporate history. http://www.research.philips.com.

Sony Group. "Chasing and Being Chased." *Sony History*. http://www.sony.net/Fun/SH.

Key-frame computer animation:

"The Film Animator Today: Artists Without a Canvas." *Science Dimension 3*, no. 5 (Oct. 1971).

National Research Council Canada. "Where It All Started—Animation in the NRC Labs." http://www.nrc-cnrc.gc.ca.

Tom Spears. "Animation Pioneers to Get Academy Award: Computer Technique Got Early Start at Ottawa's NRC." *Ottawa Citizen*, Jan. 9, 1997.

Waffle-sole running shoes:

Richard Goldstein. "Bill Bowerman, 88, Nike Co-Founder, Dies." *New York Times*, Dec. 27, 1999.

Nike Inc. Corporate history. http://www.nikebiz.com/company_overview/history.

"Swift Profits: Nike Shoes are Frontrunners." *Time*, June 20, 1980.

Mountain bike:

Breezer Bikes. "Joe Breeze and the History of Breezer Bikes." http://www.breezerbikes.com.

Editors of *Bicycling* magazine. *The Noblest Invention: An Illustrated History of the Bicycle*. Rodale Press, 2003.

Gary Fisher Mountain Bikes. "The Gary Fisher Story." http://www.fisherbikes.com.

Dan Koeppel. "Joe Breeze Wants to Change the World." *Bicycling* 44, no. 8 (Sept. 2003).

Matt Phillips. "The History of the Revolution." *Mountain Bike* 17, no. 8 (Aug. 2001).

Digital video recorder:

Mariam Fam. "Competition, Technology Enhance the DVR; TiVo, Telecom Companies Vie For TV-Recording Consumers; Heartburn on Madison Avenue." *Wall Street Journal*, March 16, 2007.

Katie Hafner. "Now Preening on the Coffee Table: The TiVo Remote Control." *New York Times*, Feb. 19, 2004.

Michael Lewis. "Boom Box." *New York Times Magazine*, Aug. 13, 2000.

Julio Vasconcellos and Matt Wyndowe. "Mike Ramsay, Founder of TiVo." Iinovate podcast, Sept. 6, 2006.

CHAPTER 03: COMMUNICATION

Answering machine:

Electronic Tele-Communications Inc. Corporate history. http://www.etcia.com/history.html.

Douglas Martin. "Joseph J. Zimmermann Jr., 92, an Inventor, Is Dead." *New York Times*, April 11, 2004.

David Morton. "Answering Machines and the U.S. Telephone Companies." *Recording History: The History of Recording Technology.* www.recording-history.org.

Radio-frequency identification:

"The History of RFID Technology," RFID Journal. http://www.rfidjournal.com.

Jeremy Landt. "Shrouds of Time: The History of RFID." *Association for Automatic Identification and Data Capture Technologies*, 2001.

Melanie Rieback, Bruno Crispo, and Andrew Tanenbaum. "The evolution of RFID security." *IEEE Pervasive Computing* 5, no. 1 (Jan.–March 2006): 62–69.

Harry Stockman. "Communication by Means of Reflected Power." *Proceedings of the Institute of Radio Engineers*, Oct. 1948.

Bar code:

Russ Adams. "Bar Code History Page." Adams Communications.

Charles Fishman. "The killer app—bar none." *Fast Company* 47 (June 2001).

Leslie Goff. "Grocery Scanners Check In." *Computerworld* 33, no. 22 (May 31, 1999).

Deborah Kreuze. "Interfaces: The century's top 10." *Technology Review* 102, no. 6 (Nov.–Dec. 1999).

The Kroger Co. corporate history. www.thekrogerco.com.

George Laurer. "Development of the U.P.C. Symbol." Raleigh, North Carolina, 2001.

Steve Lohr. "Bar Code Détente: U.S. Finally Adds One More Digit." *New York Times*, July 12, 2004.

Tom Reynolds. "A Short History of Barcode Scanners." National Barcode. http://www.nationalbarcode.com.

Nicholas Varchaver. "Scanning the Globe." *Fortune* 149, no. 11 (May 31, 2004).

Magnetic stripe card:

Reed Albergotti. "Plastic Puts Fresh Crease in Cash: Debit, Credit Cards Stand as the Preferred Methods to Pay for In-Store Sales." *Wall Street Journal,* Dec. 16, 2003.

"Alumni Distinguished Service: Forrest Parry." *SUU in View*, Fall 2004.

Steve Halliday. "Historical Overview of the Card Industry." *Cardtech/Securtech*, April 27, 1998.

Peter K. Sheerin. "Bill Wattenburg's Background: Magnetic Stripe Cards." http://www.pushback.com, 2004.

Jerome Svigals. "35th Anniversary of the Introduction of the Magnetic Striped Transaction Card." International Card Manufacturers Association, 2005.

Visa Inc. "VisaNet Fact Sheet." Oct. 2007. http://corporate.visa.com.

Communications satellite:

"Another Satellite to Replace Wayward Pager Relay." CNN, May 20, 1998.

John R. Pierce. "ECHO: America's First Communications Satellite." SMEC *Vintage Electric* 2, no. 1 (1990).

John R. Pierce. "Telstar, a History." SMEC *Vintage Electric* 2, no. 1 (1990).

Roger Smith. "Communications satellite." In *Inventions and Inventors*. Salem Press, 2002.

"Telstar's Triumph." *Time*, July 20, 1962.

Calvin Tomkins. "Profile of Dr. John R. Pierce." *The New Yorker*, Sept. 21, 1963.

Touch-Tone dialing:

Bell Telephone Laboratories. "The Touch-Tone 'Dial.'" *Report from Bell Telephone Laboratories*, March 1964.

R. L. Deininger. "Human Factors Engineering Studies of the Design and Use of Pushbutton Telephone Sets." *The Bell System Technical Journal* 39, no. 4 (July 1960).

Alex Hutchinson. "Why do the numbers on my phone keypad Go Down from 1-to-9 but Are the Other Way Around on My Computer Keyboard?" *Ottawa Citizen*, July 2, 2006.

"Yet Another First Step," *The New Yorker*, April 8, 1967.

Fiber optics:

Corning Inc. "Fiber 101." http://www.corning.com.

David R. Goff. *Fiber Optic Reference Guide: A Practical Guide to Communications Technology*. Focal Press, 2002.

Jeff Hecht. *City of Light: The Story of Fiber Optics*. Oxford University Press, 1999.

Donald B. Keck. "Optical Fiber Spans 30 Years." *Lightwave*, July 2000.

Electronic mail:

Sasha Cavender. "Legends: Ray Tomlinson." *Forbes*, Oct. 5, 1998.

"Email Home." site maintained by Ray Tomlinson. http://www.openmap.bbn.com/-tomlinso.

John Naughton. *A Brief History of the Future: The Origins of the Internet*. Diane Publishing Co., 2000.

Michael Specter. "Damn Spam." *The New Yorker*, Aug. 6, 2007.

Brad Templeton. "Reflections of the 25th Anniversary of Spam." http://www.templetons.com/brad.

Cell phone:

AT&T Inc. "1946: First Mobile Telephone Call." AT&T corporate history, http://www.att.com.

Consumer Electronics Association. *Digital America* 2006. http://www.ce.org.

Joel Engel. "Joel Engel: An Interview Conducted by David Hochfelder." Transcript, IEEE History Center, Rutgers University, Sept. 30, 1999.

Ted Oehmke. "Cell Phones Ruin the Opera? Meet the Culprit." *New York Times*, Jan. 6, 2000.

"Public Got First Cell Phone Demo 30 Years Ago." *Associated Press*, April 3, 2003.

Mobile e-mail:

Louis E. Frenzel. "The BlackBerry: Reaps the Fruits of Innovation." *Electronic Design* 52, no. 7 (March 29, 2004).

"The Innovation Specialists." *Time*, July 11, 2004.

"The Interview: Mike Lazaridis, Research in Motion." *EE Times Great Minds, Great Ideas Project*, 2006.

John Markoff. "In Silicon Valley, a Man Without a Patent." *New York Times*, April 16, 2006.

Alexandra Zabjek. "How Did the BlackBerry Get Its Name?" *Ottawa Citizen*, Nov. 5, 2006.

CHAPTER 04: BIOLOGY

Carbon dating:

Willard F. Libby. "Radiocarbon Dating." Nobel lecture, Dec. 12, 1960.

Greg Marlowe. "Year One: Radiocarbon Dating and American Archaeology, 1947–1948." *American Antiquity* 64, no. 1 (Jan. 1999).

R. E. Taylor. "Fifty Years of Radiocarbon Dating." *American Scientist* 88, no. 1 (Jan.–Feb. 2000).

High-yield rice:

Lester R. Brown. *Outgrowing the Earth*. Earthscan, 2005.

Gordon Conway. *The Doubly Green Revolution: Food for All in the Twenty-first Century*. Cornell University Press, 1999.

E. J. Kahn, Jr. "The Staffs of Life." *The New Yorker*, March 4, 1985.

Frances Moore Lappe, Joseph Collins, and Peter Rosset, eds. *World Hunger: 12 Myths*. Earthscan, 1998.

World Food Prize, 2004. "World Food Prize Laureate: Yuan Longping." http://www.worldfoodprize.org.

J. C. Waterlow et al., eds. *Feeding a World Population of More Than Eight Billion People: A Challenge to Science*. Oxford University Press, 1998.

Sucralose:

Burkhard Bilger. "The Search for Sweet." *The New Yorker*, May 22, 2006.

Leslie Kaufman. "Michael Sveda, the Inventor of Cyclamates, Dies at 87."
New York Times, Aug. 21, 1999.

Lewis Stegink and Lloyd Filer. *Aspartame: Physiology and Biochemistry.* 1984.

"Sweeten to Taste." *Time*, June 5, 1950.

In vitro fertilization:

Robert E. Adler. *Medical Firsts: From Hippocrates to the Human Genome.*
Wiley, 2004.

Jane Warren. *Daily Express* interview with Louise Brown, Sept. 7, 2004.

Genetic engineering:

Harold Evans. *They Made America: From the Steam Engine to the Search Engine.*
Little, Brown & Co., 2004.

Genentech Inc. Corporate history. www.gene.com.

Genentech Inc. "First Successful Laboratory Production of Human Insulin
Announced." Press release, Sept. 6, 1978.

Nancy Rockafellar, ed. "Herbert W. Boyer Interview." University of California
at San Francisco Oral History Project, 2007.

DNA fingerprinting:

Giles Newton. "Discovering DNA Fingerprinting." *The Human Genome*,
Wellcome Trust, Feb. 4, 2004.

Giles Newton. "DNA Fingerprinting Enters Society." *The Human Genome*,
Wellcome Trust, Feb. 12, 2004.

U.S. Department of Energy, Office of Science. "The Science Behind the
Human Genome Project." http://genomics.energy.gov.

Polymerase chain reaction:

Robert E. Adler. "Oldest DNA Exposes Ancient Ecosystems." *New Scientist*,
April 17, 2003.

Kary Mullis. "The Unusual Origins of the Polymerase Chain Reaction."
   Scientific American, April 1990.

William Speed Weed. "Nobel Dude." March 29, 2000. http://www.salon.com.

Dolly the Sheep:

Gina Kolata. "First Mammal Clone Dies; Dolly Made Science History." *New
   York Times*, Feb. 15, 2003.

Maryann Mott. "Ten Years after Dolly, No Human Clones, But a Barnyard of
   Copies." *National Geographic News*, July 5, 2006.

J. Madeleine Nash. "Dr. Ian Wilmut . . . and Dolly." *Time*, Dec. 29, 1997.

Wade Roush. "Genetic Savings and Clone: No Pet Project." *Technology Review*,
   March 2005.

Ian Sample. "Who Really Made Dolly?" *The Guardian*, March 11, 2006.

Human Genome Project:

Robert E. Adler. *Medical Firsts: From Hippocrates to the Human Genome.*
   Wiley, 2004.

Leslie Roberts. "Controversial from the Start." *Science*, Feb. 16, 2001.

J. Craig Venter. Prepared statement of J. Craig Venter before the
   Subcommittee on Energy and Environment, U.S. House of Representatives
   Committee on Science, April 6, 2000.

Gerald Weissman. *Year of the Genome.* Times Books, 2003.

CHAPTER 05: CONVENIENCE

Instant camera:

Mike Brewster. "Instant Photos, Lasting Fame." *Business Week*, June 29, 2004.

Harold Evans. *They Made America: From the Steam Engine to the Search Engine.*
   Little, Brown & Co., 2004.

Victor K. McElheny. "Edwin Land." *Biographical Memoirs*, vol. 77. National
   Academies Press, 1999.

Polaroid. Corporate history. http://www.polaroid.com.

Credit card:

Credit Card Franchise Services. Report by the United Kingdom Secretary of
State for Trade, Sept. 1980.

Visa Inc. Corporate history. http://www.visa.com.

Timothy Wolters. "'Carry your credit in your pocket': The Early History of
the Credit Card at Bank of America and Chase Manhattan." *Enterprise &
Society*, June 2000.

Photocopier:

John Brooks. "Profiles: Xerox." *The New Yorker*, April 1, 1967.

David Owen. Copies in Seconds: *How a Lone Inventor and an Unknown
Company Created the Biggest Communication Breakthrough Since Gutenberg—
Chester Carlson and the Birth of the Xerox Machine*. Simon & Schuster, 2004.

Soft contact lenses:

Bausch & Lomb. "The Bausch & Lomb Story." http://www.bausch.com.
"Contact." *The New Yorker*, June 22, 1940.

Michael T. Kaufman. "Otto Wichterle, 84, Chemist Who Made First Soft
Contacts." *New York Times*, Aug. 19, 1998.

Timothy McMahon et al. "Twenty-five Years of Contact Lenses: The Impact
on the Cornea and Ophthalmic Practice." *Cornea*. 19, no. 5 (Sept. 2000).

"Otto Wichterle (1913–98)." *Nature* 395, Sept. 24, 1998.

Jan Sebenda and Milos Hudlicky. "Otto Wichterle (1913–1998): The Father of
Soft Contact Lenses." *Journal of Polymer Science: Part A: Polymer Chemistry*
37 (1999).

Bob Tedeschi. "E-Commerce Report." *New York Times*, June 21, 2004.

Jessica Whiteside. "How Were Contact Lenses Invented?" *Ask Us*, University
of Toronto.

Quartz watch:

"Electronic Quartz Wristwatch, 1969," IEEE History Center, 2004.

George Rostky. "In the '70s, Time Was of the Essence." *Electronic Engineering
Times*, Feb. 24, 1997.

Seiko Epson Corp. "Seiko Quartz Astron 35SQ." http://www.epson.co.jp/e.

Automated teller machine:

"ATM Fact Sheet," American Bankers Association, 2007. http://www.aba.com.

Robert Cole. "Personal Finance: Money Dispenser." *New York Times*, July 16, 1970.

Ellen Florian. "The Money Machines." *Fortune*, July 26, 2004.

Michael Lamm. "Dollars Ex Machina." *Invention & Technology Magazine*. 16, no. 1 (Summer 2000).

Brian Milligan. "The Man Who Invented the Cash Machine." *BBC News*, June 25, 2007.

Don Wetzel. "Interview with Mr. Don Wetzel" by David K. Allison, National Museum of American History, Sept. 21, 1995.

Electronic calculator:

Rick Bensene. "Friden EC-130 Electronic Calculator." The Old Calculator Web Museum, 2003. http://pail.bensene.com/friden130.html.

David Cole et al. *Encyclopedia of Modern Everyday Inventions*. Greenwood Press, 2003.

Frank da Cruz. "The IBM Selective Sequence Electronic Calculator." Columbia University Computing History. http://www.columbia.edu/acis/history/ssec.html.

Kathy B. Hamrick. "The History of the Hand-Held Electronic Calculator." *American Mathematical Monthly* 103, no. 8 (Oct. 1996).

Cliff Stoll. "When Slide Rules Ruled." *Scientific American* 294, no. 5 (May 2006).

Sony Walkman:

Larry Rohter. "An Unlikely Trendsetter Made Earphones a Way of Life." *New York Times*, Dec. 17, 2005.

Sony Group. "Just Try It." *Sony History*. http://www.sony.net/Fun/SH.

MP3 player:

"Dispute Over On-Line Music Is Settled." *New York Times*, Aug. 5, 1999.

Fraunhofer Institute for Integrated Circuits. "The MP3 History," 2007.

Recording Industry Association of America. "2006 Year-End Shipment Statistics." http://www.riaa.com.

Eliot Van Buskirk. "Introducing the World's First MP3 Player." *MP3.com*, Jan. 21, 2005.

CHAPTER 06: MEDICINE

Polio vaccine:

Global Polio Eradication Initiative. http://www.polioeradication.org.

Donald Robinson. *The Miracle Finders: Stories Behind the Most Important Breakthroughs of Modern Medicine.* McKay, 1978.

Smithsonian National Museum of American History. *Whatever Happened to Polio?* Smithsonian exhibit and companion website, 2005. http://americanhistory.si.edu/polio/.

Allen B. Weisse, M.D. Medical Odysseys: *The Different and Sometimes Unexpected Pathways to 20th-Century Medical Discoveries.* Rutgers University Press, 1991.

Evelyn Zamula. "A New Challenge for Former Polio Patients." *FDA Consumer*, June 1991.

Ultrasound imaging:

Association for Medical Ultrasound. "History of the AIUM." http://www.aium.org.

A. Kurjak. "Ultrasound scanning: Prof. Ian Donald (1910–1987)." *European Journal of Obstetrics & Gynecology and Reproductive Biology.* 90 (2000).

Joseph Woo. "A Short History of the Development of Ultrasound in Obstetrics and Gynecology." 2002. http://www.ob-ultrasound.net.

Edward Yoxen. "Seeing with Sound: A Study of the Development of Medical Images." In W. Bijker et al., eds. *The Social Construction of Technological Systems.* The MIT Press, 1989.

Flexible endoscope:

Hank Black. "The Enlightening Endoscope: A Tool That Transformed GI Care." *UAB Magazine* 17, no. 4 (Fall 1997).

Endo Mitsuo. "The Historical Progress of Esophagoscopic Examination." *Journal of the Japan Broncho-Esophagological Society* 53, no. 2 (2002).

Leon Morgenstern. "Fiberoptic Endoscopy Is 50! Basil I. Hirschowitz, MD." *Surgical Innovation* 13, no. 3 (2006): 165–67.

Implantable pacemaker:

John Adam. "Making Hearts Beat." Lecture, Smithsonian Institution, *Innovative Lives*, 1999. http://invention.smithsonian.org.

American Heart Association. "Cardiovascular Procedures—Statistics." http://www.americanheart.org.

"Inventive minds," *New Scientist*. 182, no. 2443 (April 17, 2004).

Medtronic Inc. Corporate history. www.medtronic.com.

Birth control pill:

Robert E. Adler. *Medical Firsts: From Hippocrates to the Human Genome.* Wiley, 2004.

Centers for Disease Control and Prevention. "Live Births, Birth Rates, and Fertility Rates, by Race: United States, 1909–94." http://www.cdc.gov.

Suzanne White Junod. "FDA's Approval of the First Oral Contraceptive, Enovid." *Update*, July–Aug. 1998.

Suzanne White Junod and L. Marks. "Women's Trials: The Approval of the First Oral Contraceptive Pill in the United States and Great Britain." *Journal of the History of Medicine and Allied Sciences* 57, no. 2 (2002): 117–60.

"The Pill." PBS *American Experience* website, 2002. http://www.pbs.org/wgbh/amex/pill.

Donald Robinson. *The Miracle Finders: Stories Behind the Most Important Breakthroughs of Modern Medicine.* McKay,1978.

Coronary bypass surgery

Igor Konstatinov. "The First Coronary Artery Bypass Operation and Forgotten Pioneers." *Annals of Thoracic Surgery* 64, no. 5 (1997): 1522–23.

Richard L. Mueller et al. "The History of Surgery for Ischemic Heart Disease." *Annals of Thoracic Surgery* 63, no. 3 (1997): 869–78.

Eric Nagourney. "Rene Favaloro, 77, a Leader in Early Heart-Bypass Surgery." *New York Times*, Aug. 1, 2000.

Nathaniel C. Nash. "Buenos Aires Journal; The Famous Dr. Favaloro Builds His Dream Clinic." *New York Times*, Aug. 18, 1992.

CT scanner

"The Beatles' Greatest Gift . . . Is to Science." *The Whittington Hospital NHS Trust News*. Jan.–March 2005.

E. C. Beckmann. "CT scanning the early days." *The British Journal of Radiology* 79, no. 937 (2006): 5–8.

"Dr. James Ambrose: Radiologist Who Made Life Easier for Countless Patients and Physicians by Helping to Invent the CT Scanner." *The Times* (London), April 25, 2006.

Godfrey N. Hounsfield. "Computed Medical Imaging." Nobel lecture, Dec. 8, 1979.

Jeremy Pearce. "Sir Godfrey Hounsfield, 84; Helped Develop the CAT Scanner." *New York Times*, Aug. 20, 2004.

Lee F. Rogers. "'My Word, What Is That?': Hounsfield and the Triumph of Clinical Research." *American Roentgen Ray Society* 180 (2003).

Robert McG. Thomas, Jr. "Allan Cormack, 74, Nobelist Who Helped Invent CAT Scan." *New York Times*, May 9, 1998.

Cyclosporine:

Roy Calne. "History of Transplantation." Lancet 368 (2006).

Karl Heusler and Alfred Pletscher. "The Controversial Early History of Cyclosporine." *Swiss Medical Weekly* 131 (2001).

B. D. Kahan. "Cyclosporine: A Revolution in Transplantation." *Transplantation Proceedings* 31 (1999).

Dr. Hartmann Stähelin's website on the history of cyclosporine. http://www.sandimmune.ch.

T. E. Starzl. "A Tribute to Jean Borel: A Transplanter's Point of View." *Transplantation Proceedings* 31 (1999).

"Transplant Pioneers Recall Medical Milestone." NPR *Morning Edition*, Dec. 20, 2004.

Ewald R. Weibel. "Who Did What?: The Human Side of the Science Enterprise." *Swiss Medical Weekly* 131 (2001).

Magnetic resonance imaging:

Harold Evans. *They Made America: From the Steam Engine to the Search Engine.* Little, Brown & Co., 2004.

Fonar Corp. "The 2003 Nobel Prize in Medicine." http://www.fonar.com.

"July 1977: MRI Uses Fundamental Physics for Clinical Diagnosis." *American Physical Society (APS) News* 15, 7 (July 2006).

"MRI's Inside Story: Case History." *The Economist.* 369 no. 8353 (Dec. 6, 2003).

The Nobel Foundation. "The Nobel Prize in Physiology or Medicine 2003." Press release, Oct. 6, 2003.

Rick Weiss. "Prize Fight." *Smithsonian* 34, no. 9 (Dec. 2003).

Prozac:

Patricia Flip. "OSU's 2003 Alumni Fellows." *Oregon Stater*, Dec. 2003.

Eli Lilly and Co. "Prozac Makes History." http://www.prozac.com.

Connie McDougall. "The Faith of a Scientist." *Response*, Winter 1998.

Protease inhibitors:

Lawrence K. Altman. "Scientists Display Substantial Gains in AIDS Treatment." *New York Times*, July 12, 1996.

Howard Chua-Eoan. "Person of the Year: Dr. David Ho." *Time*, Dec. 30, 1996.

David Ho. "Combating AIDS: Saving Lives." Interview with David Ho by Academy of Achievement, May 23, 1998. www.achievement.org.

National Institute of Allergy and Infectious Diseases. "Protease Inhibitors." *NIAID HIV/OI Searchable Chemical Database.* http://www.niaid.nih.gov.

National Institute of General Medical Sciences. "Fact Sheet: NIGMS-Supported Structure-Based Drug Design Saves Lives." 2007. http://www.nigms.nih.gov.

Richard C. Ogden and Charles W. Flexner, eds. *Protease Inhibitors in AIDS Therapy.* Informa Healthcare, 2001.

CHAPTER 07: TRANSPORTATION

Air bag:

Daimler-Chrysler. "Development History: An Idea That Triggered a Revolution." Press release, Oct.

Don Sherman. "The Rough Road to Air Bags." *Invention and Technology Magazine.* 11 (Summer 1995).

Robert Dan Spendlove. "Speed Bumps on the Road to Progress: How Product Liability Slows the Introduction of Beneficial Technology—An Airbag Example." *George Mason Law Review.* 13, no. 5 (2006).

Container shipping:

Harold Evans. *They Made America: From the Steam Engine to the Search Engine.* Little, Brown & Co., 2004.

Maersk Line. "Facts and Figures." http://www.maerskline.com.

Joe Nocera. "A Revolution That Came in a Box." *New York Times,* May 13, 2006.

Zeolite catalytic cracking:

W.-C. Cheng. "Environmental Fluid Catalytic Cracking Technology." *Catalysis Reviews—Science and Engineering* 40 (1998).

Charles Plank. "The invention of zeolite cracking catalysts." *Chemtech,* April 1984.

U.S. Patent 3,140,249.

Jet airliner:

Boeing Corp. Corporate history. http://www.boeing.com/history.

Chasing the Sun: The History of Commercial Aviation Seen Through the Eyes of Its Innovators. Website accompanying PBS documentary, 2001. http://www.pbs .org/kcet/chasingthesun.

Harold Evans. *They Made America: From the Steam Engine to the Search Engine.* Little, Brown & Co., 2004.

John Newhouse. "Big, Bigger, Jumbo." *The New Yorker,* June 28, 1982.

Vaclav Smil. *Transforming the Twentieth Century: Technical Innovations and Their Consequences.* Oxford University Press, 2006.

Fuel cell vehicle:

"The Fuel Cell Drives a Tractor." *Business Week*, Oct. 17, 1959.

Emilia Askario. "Ford CEO Says He's Green." *Detriot Free Press*, Oct. 31, 2001.

J. Hall and R. Kerr. "Innovation Dynamics and Environmental Technologies: The Emergence of Fuel Cell Technology." *Journal of Cleaner Production* 11 (2003).

Gerald Halpert et al. "Batteries and Fuel Cells in Space." *The Electrochemical Society Interface*, Fall 1999.

Jim Motavalli. *Forward Drive: The Race to Build "Clean" Cars for the Future*. Sierra Club Books, 2001.

"New Family of Power Producers." *Business Week*, June 27, 1959.

"Short Cut to Electrical Energy." *Business Week*, Dec. 6, 1958.

Charles Stone and Anne E. Morrison. "From Curiosity to 'Power to Change the World.'" *Solid State Ionics* 152–53 (2002).

Three-point seat belt:

Michelle Krebs. "The Man Who Buckled Up the World." *New York Times*, Jan. 24, 1999.

National Inventors Hall of Fame. "Nils Bohlin to Be Inducted into National Inventors Hall of Fame for Inventing the 3-point Safety Belt." Press release, May 16, 2002.

"Nils Bohlin, 82, Inventor of a Better Seat Belt." *New York Times*, Sept. 26, 2002.

Bypass turbofan:

Boeing Corp. Corporate history. http://www.boeing.com/history.

Harold Evans. *They Made America: From the Steam Engine to the Search Engine*. Little, Brown & Co., 2004.

John Newhouse. "Big, Bigger, Jumbo." *The New Yorker*, June 28, 1982.

Pratt & Whitney. "JT9D: Power into the 21st Century." www.pw.utc.com.

J. W. Witherspoon and H. A. Jackson. "An Introduction to the JT9D Engine." American Institute of Aeronautics and Astronomics, Commercial Aircraft Design and Operation Meeting, June 12–14, 1967.

Unmanned aerial vehicle:

"Drones Put U.S. Eyes Over North Vietnam." *Aviation Week*, July 24, 2000.

"The Firebee." Time, Nov. 27, 1964.

Kenneth Macksey. *Penguin Encyclopedia of Weapons and Military Technology*. Penguin Books, 1995.

Laurence R. Newcome. *Unmanned Aviation: A Brief History of Unmanned Aerial Vehicles*. American Institute of Aeronautics and Astronautics, 2004.

Northrop Grumman Integrated Systems. "Our Heritage." http://www.is .northropgrumman.com.

Spies That Fly. PBS Nova documentary and companion website, 2002. http:// www.pbs.org/wgbh/nova/spiesfly.

"Teledyne-Ryan AQM-34L Firebee." National Museum of the USAF, Wright-Patterson AFB, Ohio.

Catalytic converter:

"Clean-Air Buff." *Time*, May 29, 1972.

Corning Inc. "Corning Research Team Awarded National Medal of Technology." Press release, Feb. 15, 2005.

Manufacturers of Emission Controls Association. "The Catalytic Converter: Technology for Clean Air." http://www.meca.org.

U.S. Patent 3,885,977.

Computer-controlled ignition:

David Alan Grier. "Controlling the Conversation." *Computer*, Sept. 2007.

Steve Hatch. *Computerized Engine Controls*. CENGAGE Delmar Learning, 2005.

Trevor O. Jones. "Some Recent and Future Automotive Electronic Developments." *Science* 195, no. 4283 (March 18, 1977).

Marvin Tepper. *Electronic Ignition Systems*. Hayden Book Co., 1977.

Ben Watson. *Electronic Ignition: Installation, Performance Tuning, Modification*. Motorbooks International, 1994.

Hybrid-electric car:

Judy and Curtis Anderson. *Electric and Hybrid Cars: A History*. McFarland & Co., 2004.

Chester Dawson. "Takehisa Yaegashi: Proud Papa of the Prius." *Business Week*, June 20, 2005.

Peter Fairley. "Hybrids' Rising Sun." *Technology Review* 107, no. 3 (April 2004).

Micheline Maynard. "Getting to Green." *International Herald-Tribune*, Oct. 24, 2007.

U.S. Environmental Protection Agency. Energy fuel ratings. http://www.fueleconomy.gov.

CHAPTER 08: BUILDING AND MANUFACTURING

Latex paint:

Etcyl Blair. "History of Chemistry in the Dow Chemical Company." ACS Central Regional Meeting, May 18, 2006.

April S. Dougal. "Company History: The Glidden Company." *International Directory of Company Histories*.

H. A. Scholz. "History of Water-Thinned Paints." *Industrial and Engineering Chemistry* 45, no. 4 (April 1953).

Numerical control machining:

Norman Bleier. "An Industry Giant Leaves a Lasting Legacy." *Modern Applications News*, Oct. 2007.

Golden Herrin. "Industry Honors the Inventor of NC." *Modern Machine Shop*, 71, no. 7 (Dec. 1998).

"NC's Birthday Party." *Manufacturing Engineering*, 121, no. 20 (Dec. 1998).

Russ Olexa. "The Father of the Second Industrial Revolution." *Manufacturing Engineering* 127, no. 2 (Aug. 2001).

J. Francis Reintjes. *Numerical Control: Making a New Technology*. Oxford University Press, 1991.

Carbon fiber:

Jason Gorss. "High Performance Carbon Fibers." American Chemical Society, *Chemical Landmarks*, 2003. Available at: http://portal.acs.org.

Masayoshi Kamiura. "Toray Carbon Fiber Composite Materials Businesses." Toray Industries, 2005.

K. Morita et al. "Characterization of Commercially Available PAN-based Carbon Fibers." *Pure & Applied Chemistry* 58, no. 3 (1986).

Float glass:

"Float Glass." *Time*, Feb. 2, 1959.

John Holusha. "Alastair Pilkington, 75, Inventor of a Process to Make Flat Glass." *New York Times*, May 24, 1995.

http://measuringworth.com.

L. A. B. Pilkington. "Review Lecture: The Float Glass Process." *Proceedings of the Royal Society of London. Series A, Mathematical and Physical Sciences* 314, no. 1516 (Dec. 16, 1969).

Pilkington Group Ltd. "Sir Alastair Pilkington." Pilkington company biography. http://www.pilkington.com.

Cordless tools:

NASA Scientific and Technical Information. "Did NASA Invent Cordless Power Tools?" http://www.sti.nasa.gov.

David Olsen. "The Hole Story: How Black & Decker Spun a Great Idea and Made History." Black & Decker press release, 2002.

Rick Schwolsky. "Hall of Fame 2001: Robert H. Riley Jr." *Tools of the Trade*, Jan. 2001.

Industrial robot:

Bob Chuvala. "Domo Arigato Mr. Engelberger: 'Father of Robotics' Talks Technology and His Latest Ideas for Inventions." *Fairfield County Business Journal*, Sept. 4, 2006.

Barnaby Feder. "He Brought the Robot to Life." *New York Times*, March 21, 1982.

International Federation of Robotics. "2005 World Robot Market." http://www.ifr.org.

Leonard Lynn. "Japanese Robotics: Challenge and—Limited—Exemplar." *Annals of the American Academy of Political and Social Science (AAPSS)* 470 (Nov. 1983).

Paul Mickle. "1961: A Peep into the Automated Future." *The Trentonian*, 1998.

Michael E. Moran "Evolution of Robotic Arms." *Journal of Robotic Surgery* 1 (2007).

U.S. Patent 2,988,237.

Kevlar:

David E. Brown. *Inventing Modern America: From the Microwave to the Mouse.* The MIT Press, 2001.

DuPont. "'Survivors' Club' Recognizes 3000th Law Enforcement Officer Saved by a Protective Vest." Press release, March 7, 2006.

DuPont. "Technical Guide: Kevlar Aramid Fiber." http://www.dupont.com.

Workmate workbench:

Black & Decker. "Thirty Years On and Still Going Strong." Press release, April 4, 2003.

Ronald Hickman and Michael Roos. "Workmate." *Journal of the Institute— CIPA*, July 1982.

Gore-Tex:

W. L. Gore & Associates. Corporate history. http://www.gore.com.

"Inventor of the Week: Robert Gore," Lemelson-MIT Program, Sept. 2006.

Alex Ward. "An All-Weather Idea." *New York Times*, Nov. 10, 1985.

CHAPTER 09: HOUSEHOLD

Microwave:

David Cole et al. *Encyclopedia of Modern Everyday Inventions.* Greenwood Press, 2003.

Dave Fusaro. "Catching the Next Micro Wave: Market for Microwave Oven-Specific Foods." *Prepared Foods*, June 1994.

J. Carlton Gallawa. "Who Invented Microwaves?" In *The Complete Microwave Oven Service Handbook: Operation, Maintenance, Troubleshooting, and Repair.*

Prentice Hall, 2007.

"Inventor of the Week: Percy L. Spencer." Lemelson-MIT Program, May 1996.
http://www.measuringworth.com.

Don Murray. "Percy Spencer and His Itch to Know." *Reader's Digest*, Aug. 1958.

Raytheon Co. Corporate history. http://www.raytheon.com.

Bridget Travers and Jeffrey Muhr, eds. *World of Invention*. Gale Research, 1994.

Stephen Van Dulken. *Inventing the 20th Century: 100 Inventions That Shaped the World*. NYU Press, 2002.

Superglue:

Harry Coover. "Discovery of Superglue Shows Power of Pursuing the Unexplained." *Research-Technology Management*, Sept.–Oct. 2000.

"Inventor of the Week: Harry Coover." Lemelson-MIT Program, Sept. 2004.

Nonstick pan:

Anne Cooper Funderburg. "Making Teflon Stick." *Invention & Technology Magazine* 16, no. 1 (Summer 2000).

William Robbins. "Teflon Maker: Out of Frying Pan into Fame." *New York Times*, Dec. 21, 1986.

Tefal. Corporate history. http://www.tefal.com.

Velcro:

Martin Jacobs. "Unzipping Velcro." *Scientific American*, April 1996.

"New Ideas." *Time*, Sept. 8, 1958.

Velcro Industries. Corporate history. http://www.velcro.com.

Remote control:

Paul Farhi. "The Inventor Who Deserves a Sitting Ovation." *Washington Post*, Feb. 17, 2007.

"Pause to Remember the Man Behind the Remote, Robert Adler, Dead at 93." *Associated Press*, Feb. 16, 2007.

Zenith Electronics Corp. "Five Decades of Channel Surfing: History of the TV Remote Control." http://www.zenith.com.

Disposable diaper:

Malcolm Gladwell. "Smaller." *The New Yorker*, Nov. 26, 2001.

"Inventor of the Week: Marion Donovan." Lemelson–MIT Program.

Andrew C. Revkin. "Victor Mills Is Dead at 100; Father of Disposable Diapers." *New York Times*, Nov. 7, 1997.

Robert McG. Thomas, Jr. "Marion Donovan, 81, Solver of the Damp-Diaper Problem." *New York Times*, Nov. 18, 1998.

Pull-top can:

"Beer Can History." Brewery Collectibles Club of America. http://www.bcca.com.

"Dayton Reliable Tool Company." *Ohio History Central*, July 27, 2006.

David Hillman and David Gibbs. *Century Makers*. Welcome Rain, 1999.

Alfonso A. Narvaez. "E. C. Fraze, 76; Devised Pull Tab." *New York Times*, Oct. 28, 1989.

Henry Petroski. *The Evolution of Useful Things*. Vintage, 1994.

Wesley G. Rogers. "Working knowledge." *Scientific American* 275, no. 6 (Dec. 1996).

"Sta-Tab Beverage Can." *Humble Masterpieces* exhibit. Museum of Modern Art, New York, 2004.

"The Stay-on-Tab Designed by Dan Cudzik on a Reynolds Metals Co. Aluminum Can." Virginia Historical Society, 2005.

Smoke detector:

BRK Electronics. Corporate history. http://www.brk.co.uk.

"Duane Pearsall: Inventor of the First Practical Home Smoke Detector." Worcester Polytechnic Institute Presidential Medal, 2004.

Bernard Gladstone. "It Gives an Earlier Warning." *New York Times*, Sept. 5, 1971.

National Fire Protection Association. "Fact Sheet: Smoke Alarms." 2007.

U.S. Environmental Protection Agency. "Buying a Smoke Detector." 2007.

U.S. Patent 3,460,124.

Post-it note:

Bethany Halford. "Sticky Notes." *Chemical and Engineering News* 82, no. 14 (April 5, 2004).

"Sticking Around—The Post-it Note Is 20." *BBC News*, April 6, 2000.

3M Corp. Corporate history. http://www.3m.com.

CHAPTER 10: SCIENTIFIC RESEARCH

Hologram:

Jeff Hecht. "Holographic Visions Prove Hard to Reach." *Laser Focus World* 30, no. 9 (Sept. 1994).

Sean F. Johnston. "Absorbing New Subjects: Holography as an Analog of Photography." *Physics in Perspective* 8 (2006).

Nuclear power:

Idaho National Laboratory. "Fact Sheet: Experimental Breeder Reactor-I." http://www.inl.gov.

International Atomic Energy Agency. "From Obninsk Beyond: Nuclear Power Conference Looks to Future." Staff report, June 24, 2004.

Rick Michal. "Fifty Years Ago in December: Atomic Reactor EBR-I Produced First Electricity." *Nuclear News*, Nov. 2001.

Susan Stacy. *Proving the Principle: A History of the Idaho National Engineering and Environmental Laboratory*, 1949–1999. U.S. Government Printing Office, 2000.

World Nuclear Association. "Nuclear Power in the World Today." 2007.

Solar cell:

Bell Laboratories. "Bell Labs Celebrates 50th Anniversary of the Solar Cell." http://www.bell-labs.com, 2004.

John Perlin. "The Silicon Solar Cell Turns 50." National Renewable Energy Laboratory, Aug. 2004.

"UD-led Team Sets Solar Record, Joins DuPont on $100 Million Project." University of Delaware press release, July 23, 2007.

Sputnik:

Aleksander Gurshtein. "Korolev: How One Man Masterminded the Soviet Drive to Beat America to the Moon." *American Scientist* 86, no. 1 (Jan.–Feb. 1998).

Warren E. Leary. "Present for the Beginning: A Khrushchev Remembers." *New York Times*, Sept. 25, 2007.

"Sputnik's Secret History Finally Revealed." *Associated Press*, Oct. 2, 2007.

Laser:

Bell Laboratories. "The Invention of the Laser at Bell Laboratories: 1958–1998." http://www.bell-labs.com, 1998.

Kenneth Chang. "Gordon Gould, 85, Figure in Invention of the Laser, Dies." *New York Times*, Sept. 20, 2005.

"How the Laser Happened." *Berkeley Groks*, Nov. 22, 2004.

Douglas Martin. "Theodore Maiman, 79, Dies; Demonstrated First Laser." *New York Times*, May 11, 2007.

Weather satellite:

Jet Propulsion Laboratory. "TIROS." *Mission and Spacecraft Library.* http://msl.jpl.nasa.gov.

National Oceanic and Atmospheric Administration. "The Launch of TIROS I Weather Satellite." *NOAA Celebrates 200 Years*, 2007.

National Oceanic and Atmospheric Administration. "Polar-Orbiting Satellites: The First to Be Launched." NOAA Satellite and Information Service. http://www.oso.noaa.gov.

Light-emitting diode:

Julie Bosman. "Billboard Bookends for Times Square." *New York Times*, June 12, 2006.

"Inventor of the Week: Nick Holonyak, Jr.," Lemelson-MIT Program, April 2004.

Tekla S. Perry. "Red Hot." *IEEE Spectrum*, June 2003.

Otis Port. "Nick Holonyak: He Saw The Lights." *Business Week*, May 23, 2005.

Howard Wolinsky. "U. of I.'s Holonyak Out to Take Some of Edison's Luster." *Chicago Sun-Times*, Feb. 2, 2005.

Superconducting wire:

T. G. Berlincourt. "Emergence of Nb-Ti as Supermagnet Material." *Cryogenics* 27 (June 1987).

European Laboratory for Particle Physics. CMS: *The Magnet Project Technical Design Report*. May 2, 1997.

Liquid crystal display:

Dennis P. Barker. "Flat-Panel Market Size to Top Over $100B in 2007 with TFT LCD Market to Reach Over $100B in 2008, Says DisplaySearch." *Digital TV DesignLine*, Oct. 8, 2007.

George H. Heilmeier. "Liquid Crystal Displays: An Experiment in Interdisciplinary Research That Worked." *IEEE Transactions on Electron Devices* 23, no. 7 (July 1976).

Hirohisa Kawamoto. "The History of Liquid-Crystal Displays." *Proceedings of the IEEE*, 90, 4 (April 2002).

Otis Port. "Flat-Panel Pioneer." *Business Week*, Dec. 12, 2005.

James P. Rybak. "George H. Heilmeier and the LCD." *Popular Electronics* 11, no. 12 (Dec. 1994).

A. Tanzer. "The New, Improved Color Computer." *Forbes* 146, no. 2 (July 23, 1990).

Charge-coupled device:

Alcatel-Lucent. "Bell Labs' Willard Boyle and George Smith receive Draper Prize for the Development of the Charged-Coupled Device." Press release, Jan. 4, 2006.

"Canadian Inventor Wins Prestigious Engineering Award." *CBC News*, Feb. 21, 2006.

Eastman Kodak Co. "CCD Primer." May 29, 2001.

"Lifetime Achievement." PC Magazine Technical Excellence Awards, Jan. 2006.

George Rotsky. "In Search of the Inventors of the CCD." *Electronic Engineering Times*, Sept. 17, 2001.

"Willard S. Boyle." 2007. http://www.science.ca.

Global positioning system:

Aerospace Corp. "Our History: Dr. Bradford Parkinson." 2003.

Naval Research Laboratory. "President Announces Roger Easton Recipient of National Medal of Technology." Press release, Nov. 22, 2005.

Cheryl Pellerin. "United States Updates Global Positioning System Technology." *The Washington File*, Feb. 3, 2006.

Tony Phillips and Linda Voss. "Tick-Tock Atomic Clock." April 8, 2002. http://www.science.nasa.gov.

"President Clinton: Improving the Civilian Global Positioning System (GPS)." White House press release, May 1, 2000.

Radio Shack. "A Guide to the Global Positioning System (GPS)." 2004.

Scanning tunneling microscope:

Harry Goldstein. "A Beautiful Noise." *IEEE Spectrum*, May 2004.

Jessica Snyder Sachs. *Encyclopedia of Inventions*. Franklin Watts, 2001.

"Spelling IBM with an STM." *IEEE Virtual Museum*. http://www.ieee-virtual -museum.org.

Photography Credits

Page 157: Getty Images

Page 158: Courtesy of Glidden

Page 160: MIT Museum

Page 162: Science & Society Picture Library

Page 164: Courtesy of Pilkington

Page 166: Courtesy of Black & Decker

Page 167: Science & Society Picture Library

Page 169: Young Sil Lee/University of Delaware and U.S. Army Research Laboratory

Page 170: Science & Society Picture Library

Page 172: Getty Images

Page 174: Courtesy of W.L. Gore & Associates

Page 175: Courtesy of Zenith

Page 177: Courtesy of Raytheon Company

Page 179: Courtesy of Eastman Chemical

Page 181: Getty Images

Page 182: Getty Images

Page 184: Courtesy of Zenith

Page 186: Courtesy of Procter & Gamble

Page 189: Virginia Historical Society

Page 191: Courtesy of First Alert

Page 193: Jupiter Images

Page 195: Reprinted with permission of Lucent Technologies Inc./ Bell Labs

Page 196: Courtesy of Paul D. Barefoot

Page 198: Courtesy of Argonne National Laboratory

Page 200: Reprinted with permission of Lucent Technologies Inc./Bell Labs

Page 202: Getty Images

Page 204: Reprinted with permission of Lucent Technologies Inc./ Bell Labs

Page 207: NASA

Page 208: iStockphoto

Page 210: Getty Images

Page 212: David Sarnoff Library

Page 214: Reprinted with permission of Lucent Technologies Inc./Bell Labs

Page 216: AP Photo

Page 217: IBM Corporate Archives

Index